Microwave Radio Transmission
Design Guide

For a complete listing of the *Artech House Microwave Library,*
turn to the back of this book.

Microwave Radio Transmission Design Guide

Trevor Manning

Artech House
Boston • London

Library of Congress Cataloging-in-Publication Data
Manning, Trevor.
  Microwave radio transmission design guide / Trevor Manning.
      P.   cm. — (Artech House microwave library)
  Includes bibliographical references and index.
  ISBN 1-58053-031-1 (alk. paper)
  1. Microwave communication systems. 2. Radio—Transmitters and
transmission. I. Title. II. Series.
TK7876.M354 1999   99-41774
621.384'11—dc21   CIP

British Library Cataloguing in Publication Data
Manning, Trevor
  Microwave radio transmission design guide. — (Artech House
  microwave library)
  1.Radios—Transmitters and transmission—Design
  2. Microwave communication systems—Design
  1. Title
  621.3'8411

  ISBN 1-58053-031-1

International Standard Book Number: 1-58053-031-1
Library of Congress Catalog Card Number: 99-41774

10 9 8 7 6 5 4 3 2 1

Contents

Foreword

Microwave system design has often been described as an art partially supported by science. Certainly, experience is a critical factor for the success of any design; however, this perception is largely due to the lack of comprehensive literature on the subject. Designers have accumulated personal reference libraries made up of technical publications and a few textbooks that are now, for the most part, out of print. When divestiture occurred in the U.S. telecommunications industry, many publications were discontinued as the industry entered a competitive mode. Design techniques became proprietary company information.

At the same time, the microwave industry experienced an unprecedented growth due to the demands of cellular and PCS communications and many new workers entered this field. I am often asked the question, "Do you know of any good books on microwave systems?" and my answer has always been, "No, at least not any that are still available."

Trevor Manning's book fills this void admirably. Several chapters of this book, in particular, will clarify many of the misconceptions that abound in this field. The book covers each step in the design process from conceptual planning to transmission design to frequency planning and interference analysis. Each step details the requirements and, more importantly, explains how to do it and how to avoid potential problems and pitfalls. Essential equations and formulas are given, all of which can be handled with a pocket calculator. Both North American and ITU standards are presented.

The book covers the many radio types, equipment configurations, and antennas in current use. The treatment of the effect of radio characteristics on the overall performance is particularly informative.

I congratulate the author for this valuable contribution to microwave system design.

Forrest Gullett
Contract Telecommunication Engineering Ltd.
British Columbia, Canada

Preface

Fifteen years ago as an enthusiastic but inexperienced engineer I was asked to plan a low-capacity digital radio link to replace a multichannel analog link in one of the regional telecommunications networks in the electricity utility (Eskom) in South Africa. I didn't have any idea where to start. It seemed to be a relatively simple task because the existing 400-MHz link had been running successfully for over 15 years. The 1500-MHz band was designated for the new digital radio link, and I needed to determine if the same antenna sizes could be used and whether the path was still suitable. Thus began my search for material to assist in the planning of real radio routes. I found my university textbooks did not shed any light on the subject—their detailed mathematical analysis of electromagnetic wavefronts using Maxwell's equations, while intellectually stimulating, provided no guidance at all on how to design a radio link in practice. I found the various ITU standards completely unintelligible at the time and therefore muddled through my early designs by making familiar sounding statements such as "I've designed this with a 40-dB fade margin." Fortunately I was involved in the full planning cycle including design, procurement, installation, commissioning, and monitoring and therefore derived an understanding of the atmospheric effects on radio links from a practical perspective. I learned from my mistakes and the mistakes of others as well, as some of the early digital link replacements didn't work even over existing paths. The reason had not so much to do with the fact that the links being installed were digital but that the new links were in higher frequency bands, which require better path clearance, and that the bandwidths were higher, often with more complex modulation schemes. I was acutely aware of the fact that in the older analog links the monitoring was not very sophisticated. The supervisory systems were based on polling, which could easily miss fading events; and the traffic was mainly voice, which is very resilient to quality impairments. I could foresee that the new digital links that monitored every errored bit would have to be designed properly as the results of the design would be accurately tested by this monitoring system. I began a career-long study into the design of radio links to meet a defined quality level.

Before I was tasked by Eskom with designing a new 400 radio hop Synchronous Digital Hierarchy (SDH) radio system, I was very fortunate to be able to spend seven months in Europe gaining experience with radio link design. During 1991 I worked in the propagation research department of Telettra (now owned by Alcatel). I would like to thank Umberto Casiraghi and Luca Saini for the many hours of invaluable discussion we had on propagation that shaped my understanding of how to design a radio link properly. During those seven months I had invaluable experience being involved in research, having discussions with design engineers, and doing system designs during bid preparations. I also spent some time in the Andrew Corporation factory in Lochgelly (Scotland) talking to designers and witnessing the antenna manufacture and test facilities. I would particularly like to thank Chris Hills and Alan Kennedy for the insightful discussions that enhanced my understanding of antenna systems.

As part of my preparation for designing the Eskom network I visited the Microflect Inc. facility in Salem, Oregon. I would like to thank Ronald Ottosen for the excellent material he shared with me on the design of passive reflectors. This material has been included in this book. At that time I also visited Contract Engineering based in Vancouver, Canada, which produces the excellent Pathloss software I was using to design the low-capacity links. I would like to thank Forrest Gullett for the many hours of discussion on the finer aspects of radio link design. I would also like to thank him for taking the time to write the foreword to this book. It will be seen that much of the material in this book is based on the excellent ITU documentation that it produces in the various recommendations and reports. I would like to acknowledge the efforts of all the contributors and the organization itself for producing this material.

I am grateful to Eskom for the many opportunities afforded to me both before and during the SDH project. The job was both interesting and stimulating, and the radio surveys using helicopters that covered virtually the entire land mass of South Africa resulted in experiences that could easily fill the pages of another book. To be involved in designing a radio system, from driving the first stake into the ground to mark the site location through to the final link performance testing, was both rewarding and experience building.

I would like to thank Marinus Coetsee from Esmartel (Johannesburg, South Africa) for his assistance in producing some training booklets that have formed the basis for some of this material. The popularity of this series encouraged me to tackle the challenging task of writing this book.

Thanks also to Ian Gordon (DMC, Coventry) for all the hard work and the excellent job he did in producing all the diagrams for the book.

Finally, I thank my wife and children for their understanding and patience during the many hours of preparation that I spent on this undertaking.

1

Introduction

Many textbooks have been published and many excellent documents written on the subject of microwave radio, however, these are often filled with complex theory and complicated mathematical formulas that leave a radio systems planner feeling inadequate rather than equipped for the task of planning a network. Many other books cover the general aspects of design but stop short of providing any practical advice on how to actually proceed in a real design. A library full of excellent material exists in the form of ITU standards and papers, however, there is not a lot of advice on how to apply the standards. Specialist groups often handle the different aspects of telecommunications standards without always ensuring that changes are coordinated with other related standards. This book has been written as a handy planners guide that summarizes all the issues that need to be considered in designing a radio network, and provides some basic theory for a deeper understanding of the subject. Some mathematics have been included to aid understanding, however, it has been kept to a minimum. An attempt has been made to provide the reader with practical advice, based on experience in real networks rather than a comprehensive list of theoretical options that leave the systems planner with no idea on where to start.

1.1 What is Microwave Radio?

1.1.1 RF Spectrum

Microwave radio, in the context of this book, refers to point-to-point fixed links that operate in duplex mode. Duplex operation means that each radio frequency (RF) channel consists of a pair of frequencies for the transmit and receive directions, respectively. The baseband signal, which contains the user information, occupies a limited bandwidth depending on the modulation scheme used. This signal is modulated onto an RF carrier and is transmitted over the air as an electromagnetic wavefront. The electromagnetic spectrum use is shown in Table 1.1. One may note that microwave radio links cover the frequency spectrum from 300 MHz to approximately 60 GHz.

1.1.2   The International Telecommunications Union

The RF spectrum is a scarce resource such as coal or petroleum and, therefore, needs to be used wisely and conservatively. Various services such as mobile radio, satellites, broadcasting, and fixed terrestrial links must share this common spectrum.

Each service must be allowed to expand and grow without causing interference to any other service. The task of allocating and controlling the individual parts of this spectrum is the responsibility of the International Telecommunications Union (ITU). The ITU is an international standards body set up by the United Nations. The two main arms of the ITU that are of interest to a radio planner are the telecommunications agency called the ITU-T (formally CCITT) and the radiocommunications agency called ITU-R (formally CCIR). There are two main bodies within the ITU-R that carry out the task of coordinating radio frequencies. The first is the World Administrative Radio Conference

Table 1.1
Use for Electromagnetic Spectrum

Frequency	Wavelength	Application
10 kHz	30 km	Very low frequency-submarine communications
100 kHz	3 km	Longwave radio broadcasting
1 MHz	300m	AM radio broadcast
10 MHz	30m	Shortwave radio (Ionospheric)
100 MHz	3m	FM broadcast
150 MHz	2m	Mobile radio (PMR)
300 MHz	1m	UHF TV broadcast and UHF point to point microwave links
3–60 GHz	10 cm–0.5 mm	Microwave links
230 THz	1300 nm	Fibre optics
420–750 THz	400–700 nm	Visible light
1000,000 THz	300 pm	X-rays

(WARC), which is responsible for allocating the specific frequency bands to present and future services. The second is the Radio Regulations Board (RRB) (formerly the International Frequency Registration Board (IFRB)), which defines the international rules for frequency assignments within the bands set by WARC. The members of the RRB are elected and meet on a part-time basis up to four times a year, normally in Geneva. The ITU has divided the world into three regions, as shown by Figure 1.1. Region 1 includes Europe, Africa, and the Commonwealth of Independent States (CIS); region 2: North and South America; and region 3: Asia, Australia, and the Pacific.

The proliferation of new satellite services has resulted in many fixed terrestrial bands being sacrificed in the latest WARC meetings. The latest example of this is the plan to phase out fixed radio links in the lower half of the 23-GHz band for a new digital Direct-to-Home broadcasting satellite service. The exponential growth in mobile radio technologies such as Digital Enhanced Cordless Telecommunications (DECT) and Personal Communications Networks (PCN) is severely threatening link bands below 3 GHz. Table 1.2 shows the typical link bands used for microwave radio.

Recommendations specifying the individual radio channel within these link bands are made by the ITU-R.

1.2 Why Radio?

The demand for cost-effective, high-quality transmission systems is growing worldwide with the ever-increasing demand for telecommunications services. The explosive growth of data traffic from IT networks has resulted in a demand for new digital networks based on standards such as SDH, GSM, DECT, and the terrestrial trunked radio standard (TETRA).

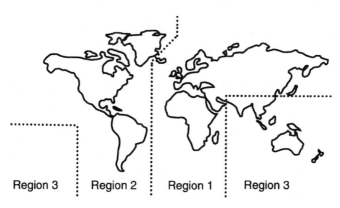

Region 3   Region 2   Region 1   Region 3

Figure 1.1 ITU regional map.

Table 1.2
Frequency Bands for Microwave Radio Links

Band	Range	Comment
2 GHz	1.7–2.7 GHz	Now designated as mobile band for PCS and DECT
4 GHz	3.8–4.2 GHz	Typically public operator band, high capacity
6 GHz	5.9–7.1 GHz	Typically public operator band, high capacity
7/8 GHz	7.1–8.5 GHz	Long haul medium to high capacity
11 GHz	10.7–11.7 GHz	Traditional public operator band, high capacity
13 GHz	12.7–13.3 GHz	Typically low to medium capacity
15 GHz	14.4–15.4 GHz	All capacities
18 GHz	17.7–19.7 GHz	Traditional public operator band, low to medium capacity
23 GHz	21.2–23.6 GHz	All capacities
26 GHz	24.5–26.5 GHz	All capacities
38 GHz	37–39.5 GHz	All capacities

The first consideration a network planner has to face is which transmission medium to use. Public Telecommunications Operators (PTOs) around the world can offer high-quality, cost-effective leased-line services in addition to satellite-based options. Fiber optic networks also require serious consideration. Fiber optic systems are very attractive for the transmission backbone due to their resilience to noise and virtually unlimited bandwidth. Ten years ago many people in the industry thought that fiber systems would, over time, completely replace radio systems. For very high capacity backbone systems this is probably true, but in the access portion of the network, radio systems are growing at an unprecedented rate. Figure 1.2 shows the growth of radio networks in the United Kingdom based on the number of licenses issued. A similar trend exists in most countries around the world for both first and third world.

Microwave radio systems are often much cheaper than satellite or leased line services and have many advantages over cable. Microwaves do not require right-of-way servitudes, which are often costly and take a prohibitively long time to obtain. Radio systems are also more cost effective at lower capacities. Cable systems have a linear cost profile—the longer the distance, the higher the cost—because most of the cost is in the cable itself. Radio systems have a stepped cost profile proportional to the number of radio repeaters required, making them less costly for longer routes. Microwave radio systems are quick and easy to install and do not impact the intervening terrain. In the case of cables, roads often have to be dug up and existing infrastructure disturbed for installation. This is costly and can have very long delays associated with it.

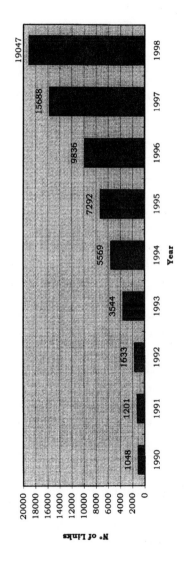

Figure 1.2 Growth of radio networks in the United Kingdom.

Many areas present almost impossible geographical challenges for cable where radio links can traverse the most difficult terrain. The success of a new network is often dependent on how quickly the network operator can offer service; new service licenses also often specify a conditional rollout rate. Radio is ideally suited to a fast rollout. Another major advantage of radio is that the equipment investment has a lower fixed cost associated with it, allowing the costs to be spread more evenly over the system's lifetime, thus reducing the investment risk. The other major advantage of microwave is that, from a return on investment (ROI) point of view, the equipment is re-usable. Microwave systems are often installed for large projects where fiber is a better long-term solution in order to take advantage of their rapid deployment; once the fiber backbone has been installed, they are decommissioned and redeployed elsewhere. In many areas microwave has been installed as a backup to fiber because it is more resilient to natural disasters such as earthquakes, nuclear disasters, and floods.

Microwave radio links often provide better overall availability since there is no risk of cable cuts, which can be very time consuming to repair. Microwave radio systems have a quicker recovery from disasters as well. Even if a tower were to fall down, the links will often work quite adequately on a path that does not have full line-of-sight (LOS) clearance, using a temporary tower structure such as the station's security fence. The background error rate of radio systems is now also comparable with fiber systems, making them an ideal transmission medium for the most challenging data services such as Asynchronous Transfer Mode (ATM).

The main limitation with radio systems is that LOS is required between sites, spectrum availability is often limited and expensive, and bandwidth is limited for very high capacity applications. In a PTO environment, fiber is popular due to the high subscriber densities that reduce the cost per subscriber profile. For access networks: GSM backhaul, wireless local loop (WLL) applications, and utilities such as electricity, gas, and water, radio is often the ideal solution.

1.3 Microwave Applications

Traditionally, microwave radio was used in high-capacity trunk routes by PTOs. These were gradually replaced by fiber optic systems that offered far higher bandwidths. Microwave radios are still used by PTOs for lower capacity portions of the network. The explosive growth in radios has been in wireless access. There are many network applications where radio is the ideal medium, as discussed earlier. Let us consider various applications for typical transmission networks.

1.3.1 Fixed-Link Operator

A fixed-link operator needs to provide backbone capacity of suitable quality and capacity to carry all the required traffic, present and future. Future expansion should be carefully considered when planning the initial installed capacity since increasing the capacity later can be costly and cause major outages. Increased capacity usually means occupying a wider bandwidth, so new frequencies need to be applied for. In some cases a new frequency band may be required with different radio equipment and antennas. Equipment that has a common design platform across all the frequency bands and transmission capacities makes this part of the planning process much easier. The network operator needs to carefully consider where traffic is likely to be added to the transmission backbone so that traffic access is catered to. The multiplexers should also be sized for future expansion. This is a balance between catering for the future and optimizing payback on capital expenditure. Most networks will require a phased approach to reduce the initial capital expenditure required in order to maximize the ROI, which is a key factor to consider in the business plan. A network operator providing a new transmission system needs to specify the number of primary circuits (e.g., E1) that have to be transported from A to B. In addition, the number of circuits to be added and dropped at each site along the route should be specified. There may also be a requirement to provide a telephone connection at the hilltop sites so that primary multiplexers are required. Typically, in a fixed-link transmission system, it is flexibility of the transmission capacity that can be accommodated and access to the network that are the key strengths of the network. Transmission quality cannot be compromised in this type of network because one needs to build the network to carry any type of traffic required. A typical carrier network is shown in Figure 1.3.

1.3.2 Utility Private Network

Utilities often use radio networks because they do not have the high-bandwidth requirements that justify fiber networks. Most utilities have private transmission networks due to the high reliability requirements dictated by the crucial circuits being carried. A power utility transmission system may need to connect a number of outlying substations or power stations to a control center. Standby control centers and regional control centers are also common, and high-quality transmission networks are required for these connections. The stability of the power grid in the country depends on the quality and reliability of the telecommunications network. VHF and UHF networks are often used for maintenance purposes, and these base stations need connections back to the various control centers. For new digital trunking networks using standards such as TETRA, these digital circuits also need to be backhauled to the main

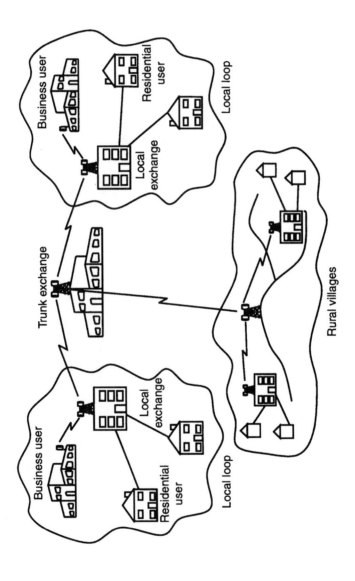

Figure 1.3 Typical carrier network diagram.

control center. In a power utility that has a meshed power grid, it is imperative to be able to disconnect the power being fed into a fault condition from the opposite end. Conventional teleprotection systems use radio signals injected onto the power line itself to switch the distant breaker off. These signals have to travel through the fault condition. It is thus desirable, where possible, to carry these teleprotection signals over a microwave radio transmission system. To avoid physical damage to the electricity transmission equipment, the microwave signal needs to trip the breaker in less than 20 ms.

These teleprotection circuits are thus very important—the consequential damage that can occur as a result of telecommunications failure can run into millions of dollars. It can thus be seen that combinations of voice and data circuits of varying importance are carried in a power utility. Similar circuits are required in most other utilities. For example, a water utility will have telemetering and telecontrol circuits from the various dams and reservoirs to report the status of water levels and control sluice gates. Utilities often carry circuits that have a very high importance yet not necessarily with high capacity. This is one of the reasons that utilities have private networks. The quality level required is thus often even higher than that provided by the public PTT. Special care should be taken when designing the network to understand the importance of the traffic irrespective of the transmission capacity. A typical utility network is shown in Figure 1.4.

1.3.3 TV Distribution Network

Transmission networks are required to connect the various radio and TV repeater stations to the main national and regional transmission sites. Radio is an attractive option because the TV broadcast sites are located on hilltops with microwave radio infrastructure such as towers and radio rooms already available. The TV signals are typically 34-Mbit/s PAL or SECAM signals with a video coder decoder (CODEC) such as Moving Pictures Export Group (MPEG-2) for digital TV. The output of the CODEC is a standard G.703 interface that can be directly connected to a radio system. In some cases video signals use 8 Mbit/s capacity. External equipment can be obtained that converts the 8-Mbit/s video signal into four E1 streams so that standard $n \times$ E1 radio transmission systems can be used. Audio channels from radio programs are often combined and compressed into a single 2-Mbit/s (E1) for transmission. TV networks often require a multidrop arrangement to drop TV circuits at various sites. A typical TV network is shown in Figure 1.5.

1.3.4 Mobile Backhaul Network

Mobile radio networks using either the GSM or DCS 1800 standard require a transmission network to backhaul the repeater traffic to the switching centers.

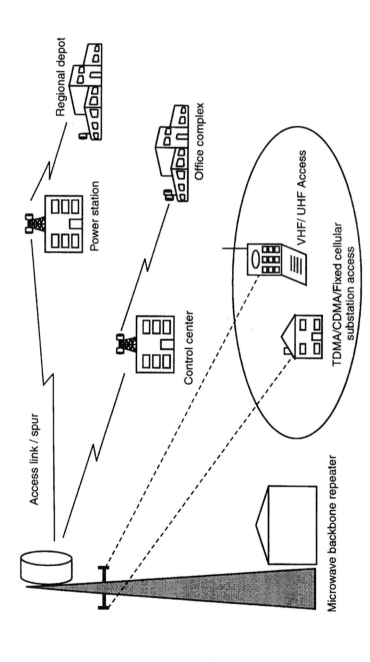

Figure 1.4 Typical utility network diagram.

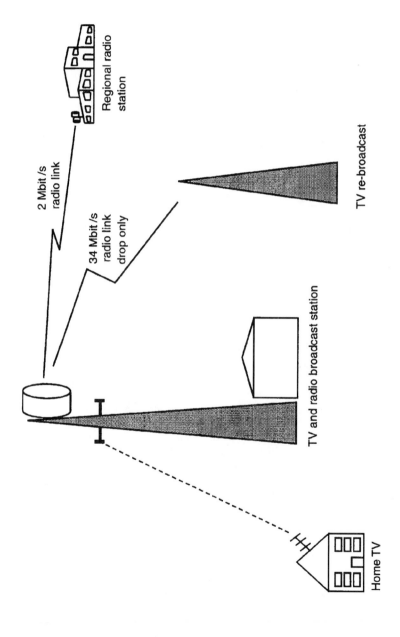

Figure 1.5 Typical TV distribution network diagram.

In a GSM environment, each base transceiver station (BTS) site usually has radios divided into a number of sectors. Each radio requires two 64-kbit/s timeslots to carry the sector traffic plus a 64-kbit/s channel for signaling. The sectors usually use subrate audio channels (16 kbit/s) using adaptive differential pulse code modulation (ADPCM), thus allowing up to four audio channels per 64-kbit/s circuit. Up to 10 transmitters can thus be accommodated on a single E1 circuit. The BTS sites can also usually carry two E1 circuits without requiring external multiplexing. This may affect the transmission arrangement. The BTS arrangement in terms of the number of sectors and radios per sector should be specified in addition to the details of the actual GSM radios. This will determine the actual capacity required to backhaul the GSM repeaters. The number of BTS sites required to be connected to the base station controllers (BSCs) and their physical location must be specified. The number of BSCs required to be connected to the mobile switching center (MSC) should also be specified. In addition, a direct public switched telephony network (PSTN) connection may be required between some of the BTSs and the MSC; the number of connections and required capacity should also be specified. A typical GSM backhaul transmission network is shown in Figure 1.6.

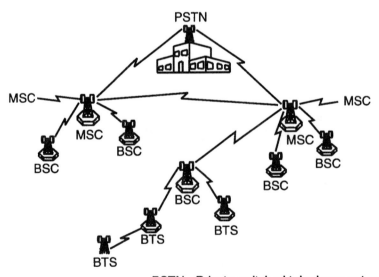

PSTN - Private switched telephone network
MSC - Mobile switching center
BSC - Base station controller
BTS -   Base transceiver system

Figure 1.6 Typical cellular backhaul network diagram.

1.4 Planning Process

Having decided that the network is to be built with radio, where does one start? Planning a radio-based network provides some interesting yet often frustrating challenges as radio waves are complex beasts that sometimes seem to have a life of their own. It is the author's view that a little bit of art needs to be combined with science to engineer a radio network, and it is this aspect that makes the work exciting and challenging. Rules based on mathematical formulae, empirical models, and complex analysis are essential for a comprehensive design but cannot replace good engineering judgment and practical experience.

System planners often start at the wrong point. A common mistake is to ask, "what fade margin should I design to?" when the correct starting point is to ascertain what the customer actually needs. The starting point is to transform the customer requirements into a transmission plan with a set of quality objectives. Fully understanding what the customer really wants and needs is one of the most difficult aspects and involves an iterative approach since the customer will often not understand the cost impact of the requirements specified. All systems experience some quality degradation and periods of failure. It is essential for the radio planner to carefully explain the quality level that can be expected and the risk of transmission failure using the selected design topology. These customer requirements need to be translated into a route diagram before the detailed planning process can begin. Detailed planning involves transforming the route plan into a detailed radio link plan, identifying suitable active and passive repeaters. A detailed radio survey is also required to plan the sites and routes. Each hop must be designed to meet certain quality objectives. Thus, these objectives need to be understood and applied. Microwave radio equipment has evolved over the past decade. A full understanding of equipment characteristics and configurations will enable the planner to choose the most appropriate equipment to meet the design objectives. Radio links suffer fading from various propagation anomalies; therefore, a thorough understanding of microwave propagation and fading mechanisms will enable the planner to design a system that is robust under the most difficult conditions. One needs to identify suitable frequency bands on which to operate the links and will have to apply to the frequency authorities for the RF channels; hence, a detailed knowledge of frequency planning and antenna characteristics is required. The final stage in the planning process—once all the aspects of design topology, quality objectives, equipment considerations, and fading mechanisms are understood—is to do the detailed link design, covering aspects such as fade margin, antenna size, and equipment protection. All these aspects are covered in detail in the following chapters. Lastly, the planner will need to make a fundamental decision for the network as to whether PDH or SDH topology should be employed. A chapter is included that discusses the two hierarchies in sufficient detail to make this decision.

2

Link Planning

Let us assume that you have been given the challenging task of performing the radio planning for a new network. Thorough planning and attention to detail in the initial stages will avoid many problems later on in the project. The infrastructure investment in a radio network is significant; therefore, the margin for error is small. Planning is an iterative process and will vary depending on the type of the installation. The planning issues in an urban area are completely different to those faced in rural greenfield sites. The link planner should identify all the steps to be followed for each site and draw up a project plan using, for example, a Gantt chart to identify and manage the critical path. A flow chart showing the link planning steps for a new radio route is shown in Figure 2.1.

2.1 Establish the Planning Brief

The very first step is to establish a planning brief. The old adage that half the solution lies in defining the problem holds true. Often a system designer is told that a $(N + 1)$-155 Mbit/s link with digital cross-connect switching is required between two sites, when in reality a far simpler and more practical solution would be possible if the true requirement was defined. The planning brief should be specified in terms of the services and related bandwidth to be carried over the transmission medium, the end-to-end circuit connections, and the required quality objective. The services to be carried could include voice circuits (300 Hz to 3400 Hz), data services at various capacities, trunk connections between public exchange switches (e.g., 2 Mb/s (E1) or 1.5 Mb/s (T1)), video circuits (e.g., 34 Mb/s (E3)), or ATM services. By understanding the true

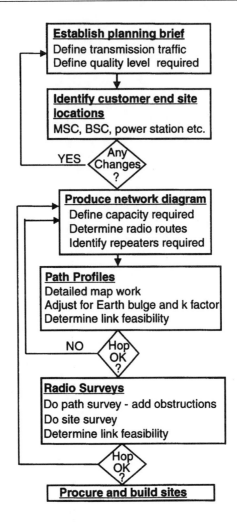

Figure 2.1 Link planning flow chart for new radio route.

number of circuits concerned and their associated bandwidths, the transmission engineer has the flexibility to design the network more efficiently. For example, a number of E1 streams that are partially filled—less than 30 circuits—can be groomed at nodal sites into a single E1 stream, thus reducing the overall transmission bandwidth required. A common error in sizing the transmission capacity required for a GSM network is to aggregate E1 circuits along the network, resulting in enormous bandwidth requirements into the MSC switch site. By understanding the true requirement, the transmission engineer has the flexibility to route circuits in different directions around the network or to build secondary routes off a main backbone route. It is essential to understand if the overall capacity is made up of lots of smaller tributaries, such as E1 circuits, or if it is a few larger tributaries such as video E3 circuits or ATM circuits. One needs to understand the real quality level required for the various circuits because some of the spur links may be implemented without protection while other spur links may be so important that two or more diverse routes may be required into the site. One should remember that the backbone has to be designed for the best quality circuit to be carried. Even if only one important circuit is to be carried on the trunk, the whole trunk circuit has to be designed to that performance standard. This is often motivation for picking up secondary, less important traffic via a secondary lower capacity route rather than taking the backbone through each site. If one has a full understanding of the overall brief, these network decisions can be made intelligently.

2.2 Initial Planning

Let us now assume that a planning brief has been specified that clearly defines what the customer is trying to achieve. The next step is to determine the initial network topology. First, the site location of the customer end sites must be determined, then an initial network diagram with the circuit connections and traffic capacity can be worked out. In practice, this will always be an iterative process since the end sites may change depending on the final position of a new building, a power station, or the area coverage of a GSM repeater. Site availability of planned repeaters is also a major variable. Switch (telephone exchange) sites, E1, E3 connections, and subrate access services should all be considered in order to understand what capacity and type of radio system is required.

2.2.1 Site Location

It is very important to check and verify information on site locations. Microwave radio links allow very little inaccuracy of the site coordinates because the clearance of the beam is critical. This will be covered in great detail later in the book. In most cases site coordinates need to be accurate to within 10m to 20m. The modern tendency is to specify site coordinates from Global Positioning System (GPS) readings. While this is a very useful and necessary tool, serious planning errors can be made if the limitations of this terminology are not clearly understood.

The GPS is a satellite-based system owned by the Department of Defense in the United States. It consists of 24 satellites in a subsynchronous orbit around the Earth. This means that, unlike the geostationary satellites that rotate at the same speed as the Earth and thus appear fixed relative to the Earth, the GPS satellites will have different positions at different times of the day. The number of satellites that will be visible at any one time will thus vary, which can affect their accuracy. The principle of the system is that each satellite sends a signal to an Earth-based portable receiver so that, using the principle of triangulation, a position fix can be made with three satellites. The additional satellites are used for error correcting to increase the accuracy. Each satellite has onboard very accurate Caesium and Rubidium clocks that are corrected by additional Earth-based clocks to achieve exceptional accuracy. Since the system was developed primarily for military use to guide missile systems, a coded jitter signal is superimposed on the reference signal from the satellites, purposefully making them inaccurate when using a standard receiver. If this selective availability (SA) is switched off, such as during wartime situations, the overall accuracy increases. Authorized military users use the Precise Positioning System (PPS), which provides excellent accuracy. For Standard Positioning Systems, the inaccuracy exceeds 100m in both horizontal and vertical planes even with maximum visibility of satellites. Inexpensive hand-held GPS systems may be significantly worse than this. In many cases, GPS systems are very useful in that qualified surveyors with expensive equipment are not required; however, it must be recognized for microwave radio applications that the more expensive site location methods will be required. These include using GPS systems in differential mode or using surveyed coordinates with survey beacons and a theodolite. Accurate differential carrier phase GPS systems can achieve an accuracy of 1 mm, however, they tend to be very expensive.

Another important consideration is to obtain the coordinates of the actual position of the antenna mounting and not just the position of the overall site. An accurate location of a building, power station, or hilltop is useless if the actual position of the antenna location is some distance from this reference point. Building new repeater sites on a major transmission route can cost hundreds of thousands of dollars; therefore, it is very important that the analysis that determines microwave radio beam clearance is accurate. In UHF and VHF systems this is not as important because the larger Fresnel zones result in lower diffraction losses. This aspect is discussed in detail later in the book. The important point to realize is that a detailed analysis of a path profile only makes sense if the site location data is accurate.

2.2.2 Network Diagram

Once the site locations have been ascertained, they should then be plotted geographically with the logical circuit connections shown to produce a network diagram. A typical GSM network showing network capacity is shown in Figure 2.2.

The network capacity can be determined by adding the number of circuits on each hop between the various end-to-end connections. Extra capacity should be added for system expansion and then the requirements are rounded off to the nearest standard bit rate.

2.2.3 Initial Mapwork

Having determined the network connections and initial capacities, a first pass radio network diagram is required. Microwave radio links rely on there being LOS between the two ends. Later in this book this aspect will be covered in great detail. As an initial planning rule though, it should be assumed that if there is no LOS (i.e., the path is blocked by the terrain itself, some trees, or a building) the path probably will not work. A repeater site will be required or the site should be moved. The first way to check for LOS, for initial planning purposes, is to do some mapwork. As a rough guide it should be assumed that radio hops should provisionally be no longer than 50-km long. In other words, if two end sites are 100 km apart, it should be assumed that one radio repeater is required roughly midpath. This assumption is purely to establish a starting point for the number of repeater sites. It should be noted that radio hops can be made to work over distances exceeding 100 km and that in some cases radio links can work without LOS, however this is the exception rather than the rule. This phase of the design is to establish the approximate number of

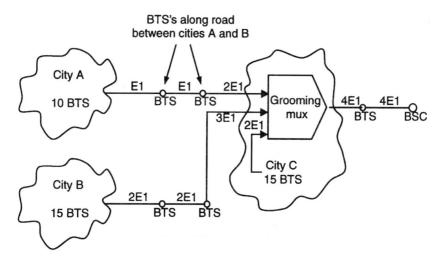

Figure 2.2 Typical GSM capacity diagram.

repeater sites required for initial costing purposes and to identify suitable routes. As in most aspects of the radio link design, the process is iterative. Good radio link design ensures that reliability of the microwave signal is balanced by a network operator's cost objectives.

2.2.4 Existing Infrastructure and Repeater Sites

2.2.4.1 Topographical Maps

A useful way to start the process of determining the radio repeater requirements is to plot the sites identified on the network connection diagram onto a geographic map. Topological maps with a scale of 1:500,000 or 1:250,000 have the right scale to provide the high-level view of the network.

Once the end sites are plotted on the map, the existing radio repeater sites can be plotted. It is also useful to get hold of other network operator's sites to be able to evaluate where site sharing is practical. Once all the sites are plotted, the possibilities can be evaluated. This high-level view will show high ground areas that may block the LOS. It also shows up areas that can be used for potential new repeater sites. The first pass assumption of 50-km hops can be used as a starting point to find suitable repeaters. In practice, suitable repeaters will seldom be found midpath, so it may be that instead of an ideal 50-km/50-km split, a site may be found with a 45-km/60-km off-path split creating a dogleg as shown in Figure 2.3.

For urban links, three-dimensional map photographs, usually on a 1:10,000 scale, can be used to identify possible repeater sites and identify possible obstructions.

2.2.4.2 Digital Terrain Model

Another very useful initial planning tool is the Digital Terrain Model (DTM)—also called Digital Elevation Model (DEM). The DTM is a national database of pixel coordinates and corresponding average pixel elevations. Typically these are based on pixel sizes of 200m by 200m resolution, however in many cases a resolution of 50m by 50m or better is available. Using this digitised information one can very quickly establish whether a LOS between two sites can be achieved. DTM databases and corresponding analysis software are in a very advanced state for area coverage predictions required for GSM and WLL networks. One needs to use this with care when establishing fixed link microwave radio routes because any error in the data could result in a very costly mistake. Pixel definitions of 200m by 200m do not provide sufficient resolution to achieve the profile accuracy required on critical paths where the first Fresnel zone may only be a few tens of meters wide. Having said this, where this

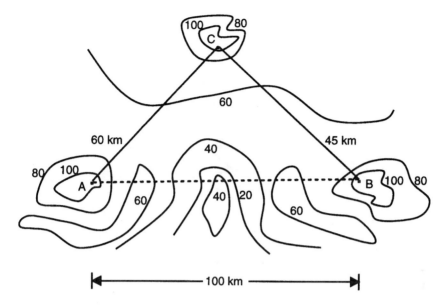

Figure 2.3 Practical path dogleg.

expensive tool is used, it can make the planning cycle considerably shorter. When used in conjunction with accurate maps, it is very useful.

2.2.5 Route Map

When planning a microwave route, various route options should be investigated. It is often a requirement from an environmental point of view that more than one route is evaluated. Site acquisition is also a crucial element in the planning process and one should keep their options as open as possible in the beginning. The output of this stage of the process should be a route map showing the recommended route and possible repeater sites, plus a number of alternative routes. These repeater sites are chosen based on the criteria of LOS considerations or where hop lengths are considered excessive. It is also crucial to include the network capacity because higher capacities typically require more complex modulation schemes to be used on the radio equipment, which limits the hop length, and thus affects the system topology. An example of a typical route map is shown in Figure 2.4.

2.3 Path Profiles

Once the overall route map has been established and possible repeater sites identified, the critical work of producing path profiles must follow. Radio

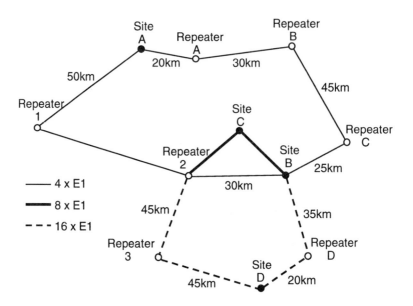

Figure 2.4 Typical route map.

repeaters, costing hundreds of thousands of dollars, are often built based on the analysis of the path profile. For long paths it is thus absolutely essential that these profiles be done accurately. In urban areas it is debatable whether a profile adds any value because terrain data will be insignificant compared to building obstructions. In these areas it is usually essential to carry out a radio survey to physically check the LOS. This is discussed in more detail later. For long rural links, mapwork analysis is essential because physically checking for optical LOS may be difficult or impossible. The preferred method of doing so is to produce a path profile, that is, a cross-section of the Earth's surface between the two end sites. Usually a 1:50,000 map is required and the contour line elevations recorded from a straight line are drawn between the two points on the map. The Earth's bulge and the curvature of the radio beam need to be taken into account. This aspect will be covered in detail later. Critical obstruction points identified on the path profile should always be physically inspected for additional obstructions such as trees or buildings.

To produce the terrain profile, the two end sites should be marked on the map and a straight line drawn between the two. Modern software programs often incorporate a map-crossing module in order to calculate the map-crossing points to avoid having to paste maps together, which can be cumbersome on long hops. The contour lines and their distances from the end site should be recorded. Due to the required accuracy, the ruler used should be carefully checked for accuracy. A steel ruler is preferred. Inexpensive plastic rulers can have inaccuracies of some millimeters, representing a few hundred meters of error.

Inexperienced planners may be tempted to mark off the profile in evenly spaced sections, say every centimeter, and then interpolate the ground height from the adjacent contour lines. This approach is not recommended because no matter how many additional points are added, no extra accuracy can be achieved. The only exact value is the contour line itself—no assumption can be made of the ground height between contours. It could be convex or concave shaped or it could be linear. Unless the ground has been physically checked, there is no way of telling. Marking off contour lines to produce a path profile is illustrated in Figure 2.5.

Usually for 1:50,000 maps, the contour lines are 20m apart in elevation. Marking off each contour can be a laborious task for hilly terrain and may add very little extra accuracy. For these types of profiles a good rule of thumb is to skip the contour lines that are evenly spaced, since they will approximate a straight line on the terrain plot anyway. One needs to be careful to incorporate spot heights and areas where the path could cross any terrain that is higher than the contour line. This is illustrated in Figure 2.6.

A worst case figure should always be used. This means increasing the value of the intervening terrain and reducing the values for the repeater sites

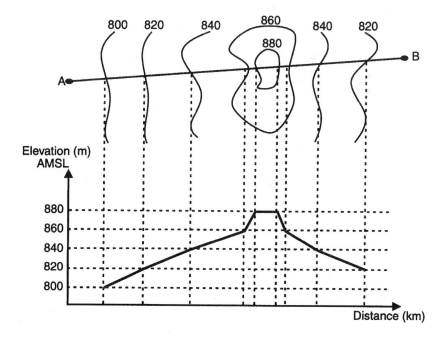

Figure 2.5 Contour map with path profile.

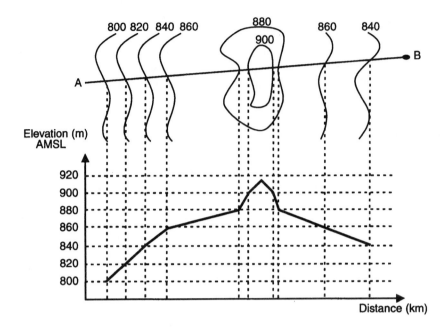

Figure 2.6 Contour map with even contours and spot height.

themselves to the highest and lowest possible values, respectively. For very flat paths it is good practice to write down a terrain value every few kilometers even though the value is constant because when the Earth bulge is calculated by a computer program it will need a base value to reference. Inexperienced planners often miss the fact that flat paths actually present the worst clearance problems due to the Earth bulge. The Earth bulge formula is [1]

$$h = d_1 \cdot d_2/12.75k \qquad (2.1)$$

where h is in meters; d_1, d_2 are the distances from each end site in kilometers, and k is the effective Earth radius factor. Using (2.1) the Earth bulge at the midpoint of a 50-km hop with a k-factor of 2/3 is

$$h = 25 \times 25/(12.75 \times 2/3) \qquad (2.a)$$
$$= 74m$$

This formula is also useful when determining what one would physically see on a radio survey. The refraction that light experiences corresponds to $k = 1.15$. Using this value of k in (2.1) will give the Earth bulge that can be expected visually, hence h (optical) = 43m.

The Earth bulge at the midpoint of a 50-km path would thus be 43m. If we assume the dominant obstacle was at this point, its height can be adjusted and one could determine if there was optical LOS or not.

Once the path profile is drawn, the preceding information can be used to adjust the profile for Earth bulge and k-factor variations. In the past, a special paper that reflected different k-factors was available, but nowadays computer programs are used to draw the modified profiles. A typical example of a path profile with modified k-factor is shown in Figure 2.7.

2.4 Radio Repeaters

The mapwork analysis and path profiles will usually highlight where new radio repeaters are required. Due to their high cost and potential delay in the project implementation, they should be carefully considered. Radio repeaters could be either active or passive.

2.4.1 Passive Repeaters

Passive repeaters are essentially "beam benders." They redirect the microwave signal around an obstruction. Passive repeaters have the following advantages over active sites:

- No power is required;
- No regular road access is required;
- No equipment housing is needed;
- They are environmentally friendly;
- Little or no maintenance is required.

All of these advantages mean that they can be built in relatively inaccessible areas. The site materials can even be transported by helicopter—the passive repeater panels are designed to be lifted by helicopter, if required. Overall service availability is not affected by site access because there is no need to visit the site once it is built.

A passive site can also be used to simplify the active repeater requirements. In other words, instead of building a tall tower with a long access road, a passive site can be used to redirect the signal to a more practical site with a short tower and short access road. This reduces cost and improves the environmental impact of the site. Since passives can be built in high areas not normally suitable

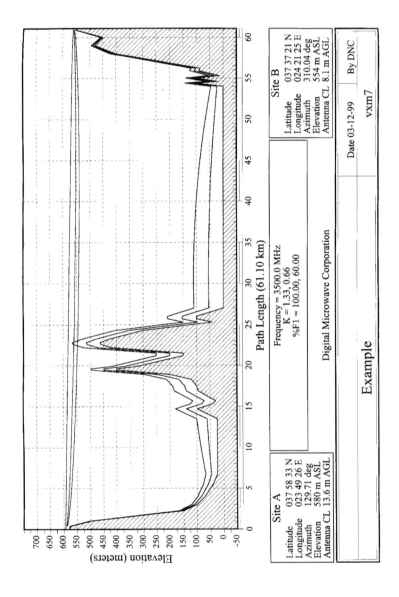

Figure 2.7 Computer-generated path profile.

for an active site, there is more flexibility to get the site to blend in with the environment.

With increasing pressure from environmental lobbyists to reduce the expansion of high sites and the exponential growth of telecomm requirements, particularly in urban areas, passive repeaters should become increasingly attractive. Urban radio links are severely hampered by LOS restrictions due to high-rise buildings. Passive repeaters, being more efficient at higher frequencies, are an ideal solution to this problem. However, although passive repeaters are very popular from an environmental and cost saving point of view, they are not so popular from a frequency planning point of view. The passive acts both as a transmit and a receive site and therefore makes frequency planning very difficult and limits frequency re-use. This tendency to waste frequency spectrum has prompted some frequency regulators to ban their use in densely populated countries such as the United Kingdom.

There are two main types of passive repeater. The first type is where two antennas are placed back to back and connected by a short feeder cable; these are called *back-to-back antenna passives*. The second type is a plane reflector type passive where a flat billboard type metal reflector is used to redirect the signal; these are often called *passive reflectors* or *plane reflectors*.

2.4.1.1 Plane Reflectors

Plane reflectors essentially consist of a large flat "drive-in screen" type aluminium plate that serves to reflect the signal and redirect it around the offending obstruction. This results in no signal distortion, because a flat conductive surface is linear. It can also support any frequency band because it is a wideband device. Being flat, large, and highly conductive also means it is 100% efficient compared to parabolic dishes that are typically only 55% efficient. They can achieve impressive gain figures due to their efficiency and the fact that they can be built to huge dimensions. Reflectors as big as 12m by 18m are readily available. The gain of passives increases with passive size. The larger the passive therefore, the greater the capture area and the greater the gain (or the less the real passive insertion loss). In order to determine the size of the passive required, one should work out a path power budget and determine the required fade margin. The required system gain should be obtained by a combination of increasing passive gain and the gain of the two antennas at the end of the link until the fade margin objective is met. Practical considerations and cost should balance an increase of antenna size and passive size. The insertion loss can be calculated as

$$IL = FSL - (FSL_1 + FSL_2) + G \qquad (2.2)$$

where *FSL* denotes overall free space loss, FSL_1 is the free space loss of the hop from site A to the passive site, FSL_2 is the free space loss of the hop from the passive site to site B, and

$$G = \text{Reflector gain (dBi)} \qquad\qquad (2.b)$$

$$= 42.8 + 40 \log f \text{(GHz)} + 20 \log A_a \text{ (m}^2) + 20 \log (\cos \theta/2)$$

where θ is the true angle between the paths.

It can be seen that one leg should be kept short in order to keep the insertion loss practical, although the huge sizes that are now available increase the possibilities.

It is strongly recommended when setting up passive reflectors that the face of the passive be lined up using survey techniques rather than panning for maximum signals. Passive reflector geometry is very critical, and therefore it is imperative that the exact coordinates be used for the calculations. Usually it will be necessary to get a surveyor to determine the exact location of the two end sites and the passive. The first step is to use these coordinates to work out the included angle (2α). This is illustrated in Figure 2.8 where the plan view is shown in the top half of the diagram and the side view is shown in the bottom half.

The azumith angles (θ_1 and θ_2) from the reflector facing to the two end sites must be calculated. The reflector vertical face angle (θ_3) can then be determined. The panning arm, which is used to rotate the passive azimuth angle by θ_3 must usually be specified when ordering the reflector to ensure it is placed at the center of its adjustment. Panning is a three-dimensional problem that has to be solved using two two-dimensional panning mechanisms. Panning, by trial-and-error only, is thus very difficult to achieve. Many frustrating days of panning such a huge structure to get the required signal levels can be avoided if the face is lined up correctly in the beginning. Fine tuning can then be achieved using trial-and-error panning.

To get the reflector face up, one needs to calculate the offset angle in each plane. To understand why there is an offset, consider the following: A passive reflector acts as a mirror to the microwave beam. According to the laws of geometric physics, the angle of incidence equals the angle of reflection. This angle is the three-dimensional angle between the three sites (two end sites plus a passive site). When translated into a two-dimensional plane, there will usually be an offset because it is unlikely that the two ends are at exactly the same height. In each two-dimensional plane therefore, the angle of incidence is NOT exactly equal to the angle of reflection. This offset angle needs to be calculated for each plane and then the face set up using a theodolite.

For environmental reasons, plane reflectors are often placed off the skyline against the mountainside. Depending on the shape of the mountain face, the

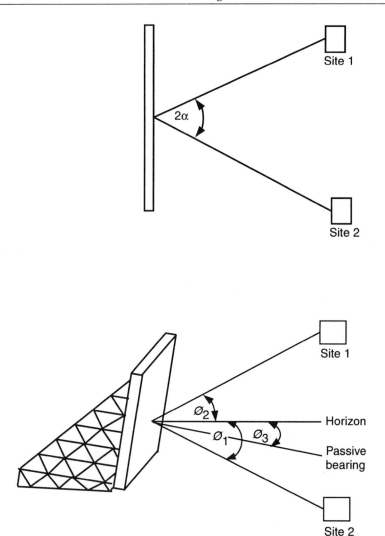

Figure 2.8 Reflector geometry.

signal can be reflected off the mountain and arrive out of phase as a delayed signal, which can result in intersymbol interference. In order for this to be a problem, the mountain must be sufficiently smooth (relative to the wavelength of the signal) and have a sufficiently large surface area, at least the area of the first Fresnel zone.

As can be seen from the preceding description, the higher the frequency the less likely reflections are to be a problem. Systems operating at 2 GHz, for example, must therefore be carefully checked, but systems operating at 8 GHz are unlikely to be affected.

Typically, reflectors are most effective when they can be placed behind the site to ensure that the included angle (2α) is small. When the reflector needs to be inline with the path, which is when this angle is greater than 130 degrees, a double reflector is required. There are two configurations that can be used. This depends on the transfer angle of the link, as shown in Figure 2.9.

One should try to balance the (2α) horizontal angles with the site layout of the reflectors. But this is not always possible due to local site conditions. The gain of the double reflector is determined by a smaller effective area. The (2α) angles should also be kept as narrow as possible to increase the effective gain. Once again, site topology may not allow this in practice.

When laying out the reflectors on site, one should try to achieve a minimum of 15 wavelengths of separation between the incoming signal and the adjacent reflector, as shown in Figure 2.10.

Double reflectors work in a close coupled mode. This means that one does not get additional gain from the second reflector but also does not need to include the path loss between the two reflectors. With double reflectors, where the passive site is inline, overshoot must be carefully checked.

2.4.1.2 Back-to-Back Antennas

Back-to-back antenna systems can be considered for short paths where there is a physical obstruction blocking the LOS. Two antennas connected by a short waveguide connection are positioned at a point where there is full LOS between each passive antenna and the respective end sites. The concept is to capture the microwave energy, concentrate it using the passive antennas, and retransmit it around the obstruction. A fundamental concept to understand when designing these systems is that the insertion loss is huge. Although one speaks of passive gain, the passive site always introduces considerable loss.

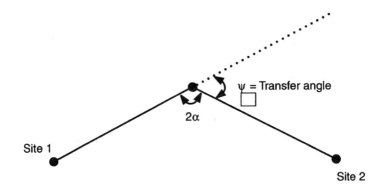

Figure 2.9 Path geometry for double reflectors.

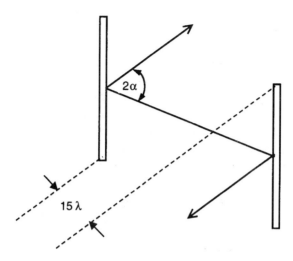

Figure 2.10 Double reflector separation.

As with plane passive sites, if the repeater is in the far-field of the two end-site antennas, the FSL is the product of the two FSLs rather than the sum. The two decibel losses must therefore be added together. This results in a very high overall FSL that must be overcome by the two back-to-back antenna gains. This limits the application to very short paths. Back-to-back antenna systems are less effective than plane parabolic passives because they are limited by the physical size of commercially available antennas and, being parabolic, are only 55% efficient. Since the path lengths tend to be very short, the main design consideration is just to achieve a useable receive signal with a minimum fade margin to ensure an adequate residual bit error ratio (RBER).

The simplest method for design calculations is to work out an insertion loss due to the passive. The insertion loss can be calculated from (2.2) replacing G with the passive antenna gains

$$IL = FSL - (FSL_1 + FSL_2) + 2A_e \qquad (2.3)$$

where *FSL* denotes the overall free space loss, FSL_1 is the free space loss of the hop from site A to the passive site, FSL_2 is the free space loss of the hop from the passive site to site B, and A_e is the antenna gain (dBi) of each passive antenna.

The advantage of this approach is that the normal path budget calculations can be done using the usual point-to-point planning software but just including a fixed attenuation equal to the value of the insertion loss. Most software planning tools allow one to insert a fixed attenuation value such as transmitter software attenuation and diffraction loss. This field can thus be used for the passive loss.

The most serious complication with passive sites is overshoot interference. Two paths develop with back-to-back systems: the direct diffracted path and the path via the antenna system. Since the passive can be installed inline, overshoot must always be considered. At face value it may appear that, provided the two legs are operated cross-polar, the interference overshoot path can be ignored. Boresight cross-polar discrimination is usually between 30 dB and 40 dB, and 30 dB to 40 dB is more than adequate for the carrier-to-interference (C/I) ratio. The error in this argument is that one is considering two separate paths. The carrier signal has traveled via the back-to-back system with an enormous extra FSL attenuation. The interference signal has only experienced the diffraction loss with no additional FSL. In addition, because the paths are normally short with oversized antennas, the signal strength is often very high. This is illustrated in the frequency section of this book. To reduce overshoot problems, the maximum gain must be placed in the back-to-back antennas. Increasing the gain of the end-site antennas will not improve the C/I ratio.

2.4.1.3 Practical Advice

Passive antenna systems can be made to work despite interference problems by taking a few design precautions. First, the overshoot path must be analyzed under the worst case k interference scenario. Remember that most paths are drawn up assuming that good clearance is desired; therefore, the end sites are often assumed to be lower than they really are, as a worst case assumption. This needs to be adjusted to the maximum worst case value for interference. Software programs may often use a low-k ($k = 2/3$) default for the worst case scenario. A high value of k (say, $k = 5$) should be used for overshoot analysis.

2.4.2 Active Repeaters

In many cases it is not possible, practical, or allowable to use a passive repeater. In this case, an active repeater will be required. A number of factors need to be taken into account when choosing a repeater site.

2.4.2.1 Site Acquisition

The most time-consuming and risky portion of the planning process is site acquisition. On a greenfield site the land owner must be contacted in order to obtain permission to build a site. This can involve complications in terms of limitations specified in the title deeds, restrictions specified by the site owner, or restrictions specified in the will of the original owner. Legal advice should be sought to check the conditions carefully. In the case of a building, the lease agreement needs to be checked. Site leasing of key sites, especially in cities, can be very lucrative to the owner. The cost of leasing a site can thus be a significant portion of the overall project; and in some cases the site may be prohibitively expensive, forcing the planner to abandon the site choice in favor of a cheaper alternative. To avoid delays, budgetary costs should be obtained early in the process.

Depending on the site infrastructure required, planning permission may have to be obtained from local councils. This could cause significant delays if public opinion has to be ascertained and can even result in a stoppage of the site plans. Even seemingly insignificant structures such as a small antenna on the side of a building may be rejected for aesthetic reasons.

2.4.2.2 Tower Issues

New sites often involve building a new tower or antenna mount structure that have various civil engineering implications. In the case of a new tower, the ground type needs to be assessed for the tower foundations. These geotechnical tests should be carried out as early as possible since the results could affect the tower costs significantly. In some cases it may lead to the site being deemed unsuitable.

In the case of an antenna mount on the top of a building, the roof structure needs to be checked for strength. The tower structure needs to be built to survive wind speeds exceeding 200 km/hr and should be rigid enough so that the tower does not twist or sway excessively in strong winds. Even in the case of an existing tower, the tower needs to be assessed for strength with the extra antenna load. There are very few standards to quantify this [2]. A good rule of thumb is to ensure the twist results in less than 10-dB deflection in a 1 in 10-year wind and survival in a 1 in 50-year wind.

Before a tower can be erected, permission must normally be obtained from the civil aviation authority. This is often irrespective of tower height. In other words, there is no minimum height below which permission is not required. Any tower that protrudes above the mean height of the runway needs to be considered by the aviation authorities. Within a radius of a few kilometers of the airport and within a few degrees around each runway, stricter conditions will exist. The aviation authorities will specify the maximum tower height allowed and whether the tower should be lit and painted. Painting of the tower must be done using the international orange and white bands. When applying for permission, one should also supply them with any relevant local conditions. For example, if a 20-m tower is being planned next to a 30-m grain silo, it is unlikely that it will need to be painted. The civil aviation authority would not normally have this level of detail in their database and therefore it should be clearly specified with the application.

There are usually no servitude rights on microwave radio paths. If a new structure blocks an existing route, the incumbent operator has no legal recourse. Despite this, it is good practice to check with existing operators whether a new tower would obstruct their LOS. On a hilltop site a new tower should be carefully placed to avoid obstructing existing links.

2.4.2.3 Roads

Provision of a suitable road is often one of the most expensive components of a site, especially in terms of the life-cycle maintenance; therefore, sites should be chosen as close to a well-maintained road as possible. Sometimes this is limited by road ordinances that stipulate a certain distance from the main road in which one cannot build. One also needs to consider the point of road access to the site. Sometimes a site can be located very close to a main road but road access from the main road is only allowed some distance away. One should try to position the road access at an existing farm access road or alongside a section of the road where overtaking is allowed in order to reduce this problem. Sections of road with a solid white line or on sharp bends will seldom result in a successful road access application. Concrete roads are a relatively maintenance-free surface for sites but tend to be expensive. Tarred roads are significantly cheaper to build but tend to dry out and break up due to the low volume of traffic on them. A good compromise is to use Hyson cells—small diamond-shaped plastic holders filled with concrete—which have been successfully used at radio sites at a fraction of the cost. These are significantly cheaper than traditional concrete strip methods and are easy to construct. The cells allow expansion and contraction to take place without breaking up the concrete.

2.4.2.4 Power

The most practical source of power for a radio site is the national low voltage (e.g., 11 kV, 22 kV) reticulation supply. A 50-KVA transformer is sufficient for most thick-route applications. For some radio equipment, solar power may be used; however, with high-capacity routes using modern equipment that is very temperature sensitive, air conditioners are mandatory. The resulting current load makes solar systems impractical. Telecommunications equipment usually runs off 48-V DC power; therefore, battery supplies are normally required. For a situation where equipment is housed in an office building, DC may be impractical. In such cases, AC supply with an uninterruptable power supply (UPS) is required.

2.4.2.5 Buildings

In most cases, for medium- and high-capacity radio routes, the most practical equipment housing is a conventional brick building. For safety reasons a separate battery room must be built and one should take special note of ventilation. Extractor fans, acid drains, and protected light switches are all issues to be considered. Containers have also been successfully used where size and future expansion are not the main concerns. Containers are particularly attractive when they can be preprepared and delivered to a site with the internal installation partially complete. For cellular applications, roadside cabinets that include the equipment racks and a power supply with a battery backup are becoming more common. Space is often limited however, so in cellular applications, the future trend will be for all-outdoor radio equipment.

2.4.2.6 General

The radio site itself also needs certain considerations. Good drainage is essential to ensure that the site is not flooded during abnormal rainfall. Radio sites on hilltops can be prone to lightning damage in some countries. It is important in these cases that the site has good earthing. It is generally accepted that a low-resistance Earth connection is not as important as common bonding of all components of the site. Lightning has a broad frequency spectrum, therefore, a low impedance rather than a low-resistive value is required. On rocky hilltop sites this is often not practical, so the tower, building, and fence should all be bonded together so that any rise in ground potential from lightning does not cause a potential difference across any of the elements. Flat copper straps of at least 50 mm^2 should be used for this bonding. It is especially important to bond both sides of the site's gate post together for safety reasons. In certain areas security can be a major consideration because the value of equipment at radio sites can be significant. The amount of copper alone can be an attraction to thieves.

2.5 Radio Surveys

It is imperative when planning a radio route that once the mapwork is complete a physical survey is conducted. This is to address issues regarding the repeater site itself as outlined previously and to check the LOS. Obstructions that are not shown on the map—such as trees, buildings, and grain silos—can block the LOS and be a showstopper for the planned route. The site coordinates should also be carefully checked.

2.5.1 Path Survey

The traditional method to check LOS is to "flash the path." This involves reflecting the sun's rays off a mirror and checking for the flash at the distant end. A shaving mirror is usually sufficient to produce a decent beam. One needs to be careful when checking for a flash that one is looking at the flash from a mirror and not incidental reflections off glass in buildings and car windows, for example. It is essential to be in two-way radio or telephone contact with the person doing the flashing to get them to alternatively start and stop flashing and ensure that you are observing the correct flash of light. Flashing a path involves technique. If accurate survey equipment is not available to determine the exact position to which one is flashing, the following procedure should be followed. A spot reflection should be beamed onto the ground and then carefully raised to the horizon. Once the general direction has been determined for the distant site, the mirror should be slowly oscillated in a raster fashion in the general direction of the site. It is almost impossible to see a beam from a distant site if the person flashing the mirror is waving the mirror in all directions. Remember, the small reflection off the mirror has to be lined up with the person's eye at the distant end. There is not a huge margin of play.

For short paths a powerful spotlight can be used. More modern techniques such as infrared lamps and specially developed microwave beams can also be used. For planning in urban areas a good telescope is often quite adequate.

If the flash does not get through, then one must investigate the cause of the obstruction. Once the obstruction has been located, its height and distance can be measured using a theodolite. The limitations of the traditional methods are obvious. The sun is often obstructed. Even on a clear sunny day, one fluffy white cloud will often find its way in front of the sun at the critical moment that flashing is required. The other major problem is that often the radio towers are not yet built. Even a cultivated field will often have crops just irritatingly high enough that the flash is not visible. In some cases a cherry-picker crane, such as those used by the fire department, can be used to elevate oneself just enough to establish the LOS. In many cases, however, this is not practical.

Where economy of scale permits, it may be practical to use a helicopter. Using this technique, a radar altimeter is fitted to the undercarriage of the helicopter and using a GPS, the LOS path is accurately flown. GPS displays will indicate the cross track error (XTE), which is the horizontal distance measured orthogonally from the straight line between the two sites. Ensuring that the pilot flies slowly enough to keep the XTE at zero, the exact path can be flown within the accuracy of the GPS system. As one flies along the path, the distance from the originating site can be displayed on the GPS and the obstructions such as trees and grain silos added to the path profile. Critical obstructions can be measured with the radar altimeter by landing alongside the obstruction, setting the altimeter to zero, and then rising and hovering to the exact height of the obstruction. The exact height of the ground itself cannot be determined in this way, however, critical ground points can be measured using Survey beacons. This process can often take many hours without the aid of a helicopter because one needs three clearly sited beacons to perform the triangulation necessary for accuracy. The surveyor is often required to drive many miles to perform this measurement. With a helicopter the surveyor carrying the theodolite reflector can easily be dropped at the reference sites and the measurement carried out within minutes.

A helicopter adds a lot of accuracy to the path profile and can be quite cost effective. If one considers that each site has to be visited anyway to do the site survey, the path survey can be carried out while transporting the site survey personnel to the site. Microwave routes are often in very remote areas and site access can be quite time consuming. If one considers the full cost of a site survey team in terms of man-hour cost, travel cost, and hotel accommodation, for example, the enormous time saving can quite quickly offset the cost of a helicopter. In the author's experience of a network in South Africa, an average of eight sites could be completed in a day using this technique as compared to two sites without.

2.5.2 Site Surveys

To plan a microwave installation properly it is essential to carry out not only a path survey but also a site survey. Even where carefully maintained records are available—and most often this is not the case—site changes may have occurred. The purpose of the site visit is to establish the accuracy of the assumptions made on site readiness and to draw up a scope of work. This list details tasks to be completed before equipment installation can occur. On this visit the following aspects should also be considered.

2.5.2.1 Site Coordinates and Altitude

Nowadays handheld GPS systems are used to check site coordinates. One needs to be careful of the accuracy achievable as discussed previously. The site coordinates used for the path profiles should be carefully checked; a detailed analysis of a path profile is meaningless if the site coordinates are inaccurate. On paths where a dominant obstruction exists, the site elevation is also a key element to check. Despite the delay and added cost, it is often justifiable to get these parameters measured by a surveyor.

2.5.2.2 Existing Tower Details

The tower height should be measured to check the existing records and to ensure there is sufficient space on the tower for the new antennas. One should also check where the antenna support struts can be secured. These struts must be secured to strong and rigid portions of the tower structure. Some tower members have tensile strength only and are not suitable for antenna support. The tower itself may need to be assessed by a civil engineer for structural strength, especially for larger diameter antennas. A detailed drawing should be made of the waveguide entry to specify where any new feeders should be routed and to ensure that there is sufficient space in the existing entry. If not, a new waveguide entry will have to be installed before installation of the feeder system can proceed. In some cases, feeder gantries from the building to the tower may be required including hail-guard covers. Waveguide cannot be twisted, so the routing of the feeder must be planned to keep any bends within the bending radius of the waveguide used. One must also ensure that any aircraft warning lights will not be obscured by the antenna installation. Local obstructions to the LOS such as other towers and buildings or trees should be checked, including the impact of a new tower or tower height extension on the LOS of other users.

2.5.2.3   Earthing

Existing earthing arrangements on the tower should be checked in order to assess where the Earth straps from any new feeders should be connected.

2.5.2.4   General

While on the site survey, one should identify space on site to offload antennas and equipment. One should also verify the primary power and battery supply adequacy and plan the equipment locations in the building. In some cases, extra racks or internal gantries may be required. These must be planned in advance or they could lead to delays during radio equipment installations. High-capacity radio equipment is temperature sensitive, so in many countries air conditioning is mandatory. The power requirements for this unit and the position in the building should be carefully checked. Considerations such as airflow in the building and the position of the sun externally should influence the unit's positioning. Details of the site access should also be carefully considered. One may be surprised at the number of different people who need to visit a new site. The list includes civil contractors, surveyors, installation crews, riggers, project engineers, and managers. Access to keys to farm gates or buildings need to be obtained for everyone who needs to visit the site. Contact details of the building superintendent should be recorded. In cases such as restricted military sites or key national sites, special permits may be required. These details need to be recorded and communicated to all relevant parties. Good site directions are also essential for equipment deliveries and installation crews. Radio hilltop sites can be notoriously hard to find. Directions such as, "Turn left after the third big tree. You can't miss it!" will cause much frustration and confusion and could be costly in terms of delivery delays.

2.6 Frequency Considerations

Once the path profiles have been analyzed and the radio repeaters finalized, a detailed radio link plan can be drawn up. Appropriate frequency bands can then be chosen for the links and applications made to the regulatory authority. This is such an important aspect that an entire chapter has been dedicated to it. Link bands are usually allocated according to the service being provided and the system bandwidth required. Spectrum is very scarce, and therefore the regulator will want to allocate it as sparingly as possible. Most regulators will not allow one to operate short hops in the lower frequency bands. In fact, in some countries such as the United Kingdom, the allocated RF bands are specified in terms of the link length. A typical link length policy is shown in Table 2.1.

The success of a project can sometimes be determined by the availability of suitable frequencies. For a larger network it is therefore essential to discuss the bandwidth requirements with the regulator in advance of starting the detailed design.

Once the link designs are complete and a suitable RF band identified, an application should be made for the specific frequency pair on which each hop will be licensed. The regulator will usually require the site coordinates, the site elevations, antenna heights, and grade of service required in order to make the frequency allocation. The regulator would carry out interference calculations and issue a pair of frequencies with a designated polarization per hop. The maximum radiated power (EIRP) is usually prescribed for interference reasons.

Table 2.1
Link Length Guideline

Frequency Link Band	Maximum Distance Allowed
7 GHz	> 30 km
13/15/18 GHz	15 km to 30 km
23/26 GHz	5 km to 15 km
38 GHz	Up to 5 km

References

[1]   Freeman, Roger, *Radio System Design for Telecommunications*, New York: John Wiley and Sons, 1987.

[2]   "Structural Standards for Steel Antenna Towers and Antenna Supporting Structures," EIA/TIA 222-E, March 1991.

3

Reliability (Quality) Standards

3.1 Introduction

This chapter on quality standards has purposely been included before detailed link design is discussed because it is not possible to design a link until one is clear what the design objectives are. Just getting a radio signal to be received at a distant point is not difficult; in fact, unwanted interference signals can be received from hundreds of miles away under certain propagation conditions. The science and engineering in designing a radio link goes into the predictions of the quality that can be expected from a given radio link design.

3.2 What Do I Aim For?

Network operators usually want their systems to operate error-free 100% of the time. For the systems planner this is impossible to achieve; and if the operator is pushed for a compromise, they will usually relax their requirement to the network being error-free and available "when they need it." This does not assist in designing a real network, so usually the planner turns to the standards provided by the United Nation's standardization body—the ITU—for guidance. The problem with these standards is that they are written for international circuits that may transverse many countries and be carried over many different transmission mediums. One really needs to understand what they are trying to achieve in order to apply them sensibly. This section is intended to provide some practical guidance in order to be able to apply these standards to meet reasonable quality standards set by the operator. Since this is a book about microwave, the objectives will be discussed relative to radio objectives. These objectives are set by the radiocommunication sector of the ITU—the ITU-R. This body was formed in March 1993, when the ITU restructured its organization into three sectors and replaced the CCIR.

3.3 Hypothetical Reference Path

One of the most difficult aspects of setting quality objectives is that every real circuit connection will be different from the next. Further, a typical international circuit connection may be carried over a copper cable from the subscriber connection to the local exchange, have an interexchange connection over a broadband radio, and be linked into the international gateway exchange via a fiber optic cable and then into the other country's gateway via a satellite connection. The way the ITU have handled this complexity is to provide advice for a typical circuit connection that can then be applied to a real connection. These typical scenarios are called hypothetical reference paths [1].

The ITU-T, for example, specifies that the international hypothetical reference connection is 27,500-km long from T-reference point to T-reference point and is divided into three quality grades [2]. This is illustrated in Figure 3.1.

The ITU-R [3] assumes that the radio portion of the 27,500-km connection does not exceed 2500 km. It specifies a hypothetical reference digital path for radio-relay systems as a 2500-km path with 9 sets of multiplexing equipment for each direction and 9 consecutive identical digital radio sections of equal length (280 km). The hypothetical reference path is illustrated in Figure 3.2.

The CCIR [4] divided interruptions into those that exceed 10 sec and those that do not. The ITU-R [5] has maintained this distinction. The rationale behind this division is that long outages can be reduced by rerouting circuits

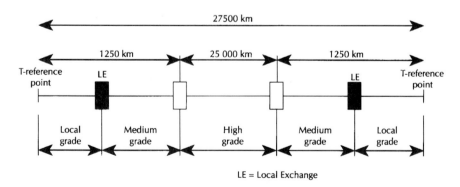

LE = Local Exchange

Figure 3.1 ITU G.821 hypothetical reference connection.

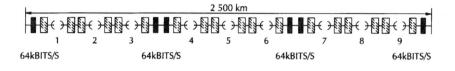

Figure 3.2 ITU-R hypothetical reference path for radio-relay systems.

on to alternative transmission systems; while with short interruptions there is insufficient time for rerouting, so other network improvement measures must be used. This distinction is crucial to understanding network protection schemes (discussed later) and is widely misunderstood in the industry. In the United States the reliability standards combine the two, however, despite recent attempts to bridge them, the main ITU standards relating to quality still require one to separate the two concepts.

For long outages, where the system is unusable for more than 10 sec, the circuit is considered to be unavailable. Unavailability standards set by the ITU limit the amount of time per annum that the circuit can be down. This includes periods in which the bit error ratio (BER) is worse than 10^{-3} and periods in which the link is disconnected.

For short outages, where the outage is less than 10 sec, the system is defined to be available (even though it is not usable by the user during this period) and during this period, performance standards are defined. This limits the amount of time per month that short outages can occur.

3.4 Unavailability Standards

Unavailability has a special meaning in the ITU standards. According to ITU-R [5] , the period of unavailable time begins when, in at least one direction of transmission, one or both of the following conditions occur for 10 consecutive seconds: either the digital signal is interrupted (i.e., alignment or timing is lost) or the BER in each second is worse than 1×10^{-3}. These 10 sec are considered part of the unavailable time.

The period of unavailable time terminates when for both directions of transmission, both of the following conditions occur for 10 consecutive seconds: the digital signal is restored (i.e., alignment or timing is recovered) and the

BER in each second is better than 1×10^{-3}. These 10 sec are considered part of available time.

3.4.1 Causes of Unavailability

Long interruptions can usually be considered in three categories:

1. Propagation;
2. Equipment failure;
3. Other.

3.4.1.1 Propagation

Outages related to propagation that last longer than 10 sec are due primarily to three causes:

1. Diffraction loss;
2. Ducting;
3. Rain.

Multipath fading is not included because the duration of most multipath fades is less than 10 sec.

Diffraction Loss

The duration of most multipath fading outages is less than 10 sec; therefore, they are considered under the performance standards. The dominant atmospheric fading effect, which affects availability, is due to diffraction of the radio signal. If the antennas are positioned insufficiently high above the ground, under certain adverse propagation conditions the radio beam will travel closer to the ground than usual, and a signal loss will occur. This loss is called a diffraction loss and occurs when a portion of the overall wavefront is obstructed by an obstacle (this will be discussed in detail later in the book). If this loss causes the receive signal to be attenuated to a level where the radio can no longer demodulate the signal, an outage will occur. In practice, by choosing suitable clearance rules, the antennas can be positioned at a suitable height so that this loss can be ignored. It is the propagation outage limit set in this section that provides the guidance for setting antenna heights.

Ducting

Ducting is a condition that can occur if the bending of the radio beam exceeds the curvature of the Earth (this is covered in detail later in the book). Under this condition, black-out fading of the signal occurs and may last several hours. In practice, this condition can usually be ignored from an outage point of view. Geographical areas that present a high risk of ducting failure are well documented. Where this condition exists, space diversity with large antenna spacing can be used to reduce its effect.

Rain

Propagation outage due to rain is proportional to the rain rate of the region. It is important to realise that it is not dependent on the average rainfall. It is the instantaneous amount of water in the path that is relevant. Water molecules absorb microwave energy by way of heating—the same principle used to heat food in a microwave oven. The greater the size of the water droplets, the greater the amount of absorption of the microwave signal. This is the reason that mist causes less rain attenuation than hard driving rain. Snow also exhibits less attenuation, although wet snow will have a greater attenuation than drizzle. One also needs to ensure that the snow or ice does not settle on the antenna, in which case the attenuation will be high. Antenna covers called radomes, heated in some cases, are often employed to ensure this does not happen. Rain attenuation causes flat fading by attenuating the receive signal. The only way to improve the availability is to increase the system gain by using, for example, larger antennas. Diversity techniques (frequency or space) provide no improvement, as both channels would be attenuated equally. Polarization diversity provides a small but insignificant improvement on the vertical polarization. The reason for this is that raindrops tend to fall as flattened droplets; thus, the attenuation in the horizontal polarization is greater than vertical polarization.

Attenuation due to rain increases as the frequency increases. Attenuation from rain attenuation is the dominant fading mechanism above 10 GHz. One does need to check the rain attenuation over long links for links just below 10 GHZ—for example, in the 8-GHZ band, since rain attenuation can exceed the margin allowed for fading, called the fade margin, in some regions even below 10 GHz. In practice, rain attenuation is so severe in the high frequency bands that it limits the length over which the radio can operate. The outage limit set in this section is the main factor that determines antenna size for links above 10 GHz.

3.4.1.2 Equipment

Long outages can occur if the radio equipment fails. The number of times the radio equipment fails is inversely proportional to the mean time before failure (MTBF) of the equipment. The duration of the outage is determined by the length of time it takes the maintenance team to restore service or the mean time to restore (MTTR). This figure includes travelling time, actual time to repair the fault, and the availability of spares. The availability (A) of a terminal is given by the formula

$$A = (\text{MTBF}/(\text{MTBF} + \text{MTTR})) \times 100\% \qquad (3.1)$$

Even for equipment with an excellent MTBF and a MTTR of a few hours, the overall availability is unacceptable for most critical networks unless route diversity or equipment protection is employed.

As an example, assume that the MTBF of a nonprotected radio terminal is 100,000 hr and the MTTR is 4 hr. The equipment link MTBF will be

$$\text{MTBF(Link)} = \text{MTBF(Terminal)}/2 \qquad (3.2)$$

Using (3.1) and (3.2) the link availability figure can be calculated as

$$A = 50,000/(50,004) \times 100 = 99.992\% \qquad (3.a)$$

The unavailability of the link is the balance of the availability from 100%; hence,

$$U = 100 - A\% \qquad (3.3)$$

The link unavailability is thus $U = 100 - 99.992\% = 0.008\%$. If one assumes a route with 20 hops, the resultant unavailability figure for the route (radio equipment outage only) is $U \times 20 = 0.16\%$. Thus, the availability of the route is

$$A \text{ (20 hop route)} = (100 - U) \qquad (3.b)$$
$$= 99.84\%$$

This figure means the route would have an outage of 14 hr/yr from equipment failure only, which for high-grade networks is totally unacceptable.

Using protected (hotstandby) radios without errorless (hitless) switching (assume 25-ms switchover time), the protected terminal availability changes to

$$A_{\text{HSB}} = \sqrt{(\text{MTBF}_A/(\text{MTBF}_A + 20 \text{ ms}) \times \text{MTBF}_B/(\text{MTBF}_B + 20 \text{ ms}))}$$
$$= \sqrt{(100,000/(100,000 + 0.02/60/60)^2} \qquad (3.4)$$
$$= (100,000/(100,000 + 0.0000055)$$
$$= 99.999999999945\%$$

These calculations show that for high-grade applications the equipment must be protected; errorless (hitless) switching, however, is not essential since even with a short switchover time, the effect on the overall outage is negligible. Hitless switching is a technique whereby the two streams are phase aligned to enable the switch to occur without any break in transmission or errors to occur on the output channel.

It is important to realize that in a Hotstandby arrangement where the two transmitters are on the same frequency, it is not possible to operate both simultaneously. A transmit switch is mandatory. On the receive side a hybrid is used to divide the signal into two paths. For an evenly split hybrid arrangement, 3-dB loss (half the power in each direction) is the best case scenario. Practically speaking, including manufacturing tolerances and extra losses, the hybrid loss is usually around 4 dB. To avoid this significant signal loss, unsymmetrical hybrids are used to divide the power so that an insignificant loss is experienced on one path with the remaining loss occurring on the standby path. A 1-dB/10-dB split is common. Network operators often have an unjustified prejudice against this arrangement, as 10 dB seems an unacceptable loss. It is important to realize that the standby channel would be seldom used as shown by the preceding calculations—it is just there to protect the path during equipment failure. All that would happen is the standby path would have a reduced fade margin. On digital systems, due to the threshold effect, one would not notice this reduced fade margin unless fading occurred simultaneously with the equipment outage. One is thus trading off, having a permanent 3-dB reduction in fade margin (at least one antenna size increase) against a reduced fade margin for the MTTR period (typically 4 to 8 hr). The former is clearly the preferred option.

3.4.1.3 Other

This category includes such events as planned maintenance outages, failure in the primary power supply, and catastrophic failure such as fire in an equipment room or the tower falling down. The only way to ensure that this type of failure does not lead to excessive outages is to have some form of route diversity in the network.

3.4.2 Unavailability Objectives

The ITU-R gives objectives relative to a hypothetical reference connection and in some cases for real links. The ITU-R objectives can be divided into the three quality grades specified by ITU-T recommendation G.821, namely, high grade, medium grade, and local grade. The ITU-T recommendation corresponds to a 27,500-km hypothetical reference circuit, 25,000 km of which is considered to be high grade and 2,500 km (1250 km at each end) that is divided between medium grade and local grade. The ITU-R [5] provides the overall availability objectives for a 2500-km digital path.

3.4.2.1 High Grade

In practice, it would normally be assumed when designing a radio system that the backbone portion of the network would be designed to high grade. The availability objectives for a real radio link forming part of a high grade circuit are specified by the ITU-R [6]. The standard specifies that the availability (A) for a high-grade circuit of link length (L) between 280 km and 2500 km should be

$$A = 100 - (0.3 \times L/2500)\% \qquad (3.5)$$

3.4.2.2 Medium Grade

There is no clear guidance on which portion should be high grade and which medium grade, for networks that are not providing PTO-type international circuits. It is usually left to the system designer to work out the relative importance of the radio link to determine which grade to apply.

The medium-grade unavailability figures are specified by the ITU-R [7]. The unavailability is given as a function of four quality classes defined for a medium-grade circuit, as shown in Table 3.1. The length of the hypothetical reference path varies with class: 280 km for classes 1 and 2 and 50 km for classes 3 and 4.

3.4.2.3 Local Grade

The local-grade specified by the ITU is used to define the quality level for a subscriber circuit—in other words, the connection from the local exchange to the subscriber terminal equipment. It could be argued that last mile type services using, for example, low-capacity point-to-multipoint topologies, could be designed to local grade. Cellular systems are generally designed to local

Table 3.1
Medium-Grade Quality Classes

Quality Class	Unavailability (%)
Class 1	0.033%
Class 2	0.05%
Class 3	0.05%
Class 4	0.1%

grade. Further guidance on availability and performance objectives for cellular systems is given by the ITU-R [8]. The type of traffic to be carried (voice versus data) and its relative importance should be carefully considered before designing a radio system to local grade; most medium- to high-capacity radio systems should be designed to high grade.

No unavailability objectives have yet been defined for the local grade, however, the ITU-R does give some guidelines [9]. The main outage effects are from equipment reliability (MTBF), repair philosophy (MTTR), and rain outage. The range of assumption being considered has led to interim values from 0.01% to 0.2% being proposed.

3.4.3 Apportionment of Objectives

The overall unavailability objective of 0.3% (high grade) should be apportioned between the three main outage categories: propagation, equipment, and other. One approach is an equal apportionment. Hence:

- Propagation (diffraction, rain, ducting): 0.1%;
- Equipment (MTBF and MTTR): 0.1%;
- Other (Maintenance and catestrophic failure, e.g.): 0.1%.

3.4.4 Practical Advice

In practice, it is very difficult for network operators to specify meaningful apportioned objectives to individual hops since most networks evolve and grow with time. This makes it very difficult for the system designer to know what availability to offer. The industry practice has thus been to interpret the objectives as follows. For propogation, the objectives are

- Rain: a link availability of 99.99%;
- Diffraction: a minimum k value specified for 99.99% (unavailability of 0.01%);
- Ducting: ignore except in high-risk areas.

In order for equipment to achieve a high-grade equipment availability objective of 99.9% over a 2500-km reference radio network, it would require unrealistically quick response times for maintenance (MTTR). To meet these objectives the equipment must be installed in a hotstandby configuration or route diversity must be possible. If this objective does not have to be met, nonprotected equipment configurations can be used.

In the other classification, there is no way to avoid most of the problems in this category other than route diversity. Therefore, in some ways the 99.9% objective is meaningless. This can be added to the rain allocation for routes that use high-frequency (e.g., 13 GHz) links.

3.5 Performance Standards

Average BER measurements—defined as the total number of bit errors divided by the total number of bits in the measurement period—is an inadequate performance metric because any long burst of errors distorts the results. A virtually error-free service will report a poor average BER if a few minutes of errors are included. The network operator could argue that if the long error burst was excluded, the average BER would be excellent. This is not an unreasonable argument since the service would in all likelihood be rerouted during the long outage period. Modern performance objectives such as G.821 and G.826 purposely exclude long bursts of errors by only defining the performance objectives during periods when the system is available. These objectives only consider one direction of transmission and are considered for a period of a month. Availability objectives, on the other hand, are considered for both directions of transmission and are measured as annual figures.

3.5.1 Causes of Outage

Short outages that affect the performance of a system are primarily caused by three affects:

1. Multipath fading effects;
2. Background errors in equipment;
3. Wind.

3.5.1.1 Multipath Fading Effects

Refraction (which is covered in detail later in this book) causes multiple radio paths to be established over the link, resulting in so-called flat fading (or more correctly Raleigh fading) outages in narrowband systems and flat and selective fading outages in wideband systems. Most of the real challenge of engineering of a radio link goes into predicting the amount of multipath fading that will occur. Many advanced techniques such as adaptive equalization are now employed in radio equipment to overcome these effects.

3.5.1.2 Background Errors

Thermal noise in radio receivers result in so-called "dribble errors." Even a fiber system that has a quality level of 10^{-13} is subject to this, as it implies that one error will occur every 10 trillion bits. In order to get similar error performance out of radio systems, forward error correction is used.

3.5.1.3 Wind

An outage mechanism that is often overlooked is wind. If the tower is not strong enough it will sway in the wind and, since the antenna beamwidth is often only a few fractions of a degree, outages can occur. In sandy areas, dust storms are often blamed for outages but the tower sway and twist is actually to blame.

3.5.2 Performance Objectives

The main ITU performance objectives are based on G.821 and G.826.

3.5.2.1 G.821 Objectives

The performance objectives are related to the following concepts as defined:

- *Errored Second* (ES): Any 1-sec period in which at least one error occurs;
- *Severely Errored Second* (SES): A 1-sec period in which the BER exceeds 10^{-3};
- *Degraded Minute* (DM): A period of 60, 1-sec periods, excluding any SES, in which the BER exceeds 10^{-6}. This period is not necessarily contiguous.

The recommendation specifies that there should be less than 0.2% of 1-sec intervals that have a BER worse than 10^{-3} (i.e., SES), less than 10% of 1-min intervals that have a BER worse than 10^{-6} (i.e., DM), and less than 8% of 1-sec intervals that have any errors (i.e., ES).

This objective specified by the ITU-T is referenced to a 27,500-km circuit connection. The ITU-R has provided recommendations based on the ITU-T one referenced to a 2500-km hypothetical reference path using radio.

High Grade

The ITU attempts to give advice for real radio links by specifying actual objectives for a real circuit in addition to the hypothetical reference connections. The hypothetical path limits and performance criteria are set by the ITU

[10, 11]. For a real high grade link of length between 280 km and 2500 km, the performance criteria are:

- BER $\geq 10^{-3}$ for no more than $0.054 \times L/2500\%$ of the worst month;
- BER $\geq 10^{-6}$ for no more than $0.4 \times L/2500\%$ of the worst month;
- ES (64 kbit/s) for no more than $0.32 \times L/2500\%$ of the worst month.

Medium Grade

For a medium-grade link, the performance objectives are specified by the ITU-R [7]. The medium-grade objectives as applied to each direction of a 64-kbit/s channel are specified relative to class 1, 2, 3, or 4. The objectives are given in Table 3.2. All figures are in percentage of any month. The length of the hypothetical reference path varies with class: 280 km for classes 1 and 2 and 50 km for classes 3 and 4.

Local Grade

For a local-grade circuit the performance objectives are defined by the ITU-R [12]. For each direction of a 64-kbit/s channel the performance objectives are defined as:

- BER $\geq 10^{-3}$ for no more than 0.015% of the worst month;
- BER $\geq 10^{-6}$ for no more than 1.5% of the worst month (measured in 1 min);
- ES for no more than 1.2% of the worst month.

The RBER is under study.

Apportionment of Objectives

The ITU-T objectives specified in standards such as G.821 and G.826 are for international connections over a 27,500-km reference circuit. Clearly it is

Table 3.2
Medium-Grade Objectives

Performance Parameter	Class 1 (280 km)	Class 2 (280 km)	Class 3 (50 km)	Class 4 (50 km)
BER > 1×10^{-3}	0.006	0.0075	0.002	0.005
BER > 1×10^{-6}	0.045	0.2	0.2	0.5
ES	0.036	0.16	0.16	0.4

important for the network designer planning a radio route of a few hundred kilometers to know the design standard so that when it forms part of the international circuit, the overall connection meets the objective. If the circuit does not form part of such a long international connection it could be argued that the entire objective could be allocated to that small network. While one would be justified in claiming adherence to the ITU standard, technically speaking the reduced quality that would be achieved would be far worse than available on other similar transmission systems, especially if optical fiber is used. The actual standard used in practice is thus a trade-off between meeting a reasonable apportioned objective and achieving a quality level using radio, that is compared to other transmission medium options that a network user would have.

The ITU-T have apportioned the three objectives of SES = 0.2%, DM = 10%, and ES = 8% over 27,500 km across the three quality grades discussed already. The high-grade portion that is assumed to be 25,000 km is allocated 40% of the overall objective; the remaining 60% is divided equally (15% each) between the medium- and local-grade portions at each end. For the high-grade portion, this results in the DM objective being 4% and the ES objective being 3.2% over 25,000 km. The ITU-R objectives assume an all-radio route with a reference length of 2500 km; hence, they allocate one-tenth of the ITU-T objective to the radio circuit, resulting in a DM objective of 0.4% and an ES objective of 0.32% The apportionment of SES is a little more complicated in that the overall objective of 0.2% over 27,500 km is divided into two 0.1% objectives: one for equipment and one for fading effects. The high-grade portion is allocated 40% of the overall equipment objective, resulting in an objective of 0.04% over 25,000 km. The ITU-R circuit of 2500 km is thus one-tenth of the objective, or 0.004%. The remaining 0.1% of the overall objective is allocated equally (0.05% each) to the medium- and high-grade portions as a block allowance (i.e., independent of distance) to account for fading. For the high-grade portion of a radio circuit over 2500 km, the SES objective is thus the sum of 0.004% plus 0.05%, or 0.054%

Practical Advice

In summary the ITU-R objectives for a 2500-km radio circuit based on G.821 are:

- SES = 0.054%;
- DM = 0.4%;
- ES = 0.32%.

It can be seen that this is consistent with the real high grade link objectives specified in the previous section. The ITU specifies that the minimum link length should be 280 km when applying these objectives; but provided each hop in the route is not very short, the actual hop length can be used to determine the design objective for the hop.

One will find that in real radio routes it is the SES objective that dominates; therefore, if one is designing a radio link to meet G.821, only this objective would be considered in determining the antenna sizes. In summary, therefore the key practical objective for the radio planner is

$$\text{Outage } (10^{-3}) < 0.054 \times L \text{ (km)}/2500\% \text{ of the worst month} \quad (3.6)$$

A two-hop link of 50 km per hop would thus be

$$\text{Outage} < 0.054\% \times (50 + 50)/2500 \quad (3.d)$$
$$= 0.00216\% \text{ of the worst month}$$

Assuming a month to have 2678,400 sec (31 days × 24 hr × 60 min × 60 sec) means that the outage in the worst month should be less than 2678,400 × 0.00216/100 = 57.85 sec/mo, which represents 29 sec (57.85/2) per month per hop of outage allowed.

3.5.2.2 G.826 Objectives

G.821 was found to be inadequate for high-capacity data services. It relies on breaking up the test period into 1-sec periods for analysis. Consider that the minimum rate specified for an SDH aggregate signal has 155 million bits in a second. It only requires one of those bits to be in error to generate an ES. Clearly the 1-sec period is too long for such systems. G.826 specifies a block of bits for analysis, which can be significantly less than 1 sec. It also specifies the objectives at the system rate rather than the circuit rate. Most practical systems run at system rates of E1, E3, and STM-1, for example, and a performance objective such as G.821, specified at E0, is not useful for practical measurements. The added complication was that, strictly speaking, G.821 is an out-of-service measurement that only makes it usable during commissioning. G.826 has been specified as an in-service measurement. The problematic DM metric has been discarded in G.826.

G.826 Definitions

- *Errored Block Ratio* (EBR): A block in which one or more bits are errored; and the block size is specified separately for each system rate.
- *Errored Second Ratio* (ESR): A 1-sec period that contains one or more errored blocks.

- *Severely Errored Second Ratio* (SESR): A 1-sec period that contains greater than 30% of errored blocks or at least one *Severely Disturbed Period* (SDP), for example, when such a loss of pointer has occurred.
- *Background Block Error* (BBE): An errored block not occurring as part of an SES.

G.826 Specification

The objectives for a 27,500-km hypothetical reference path is specified in Table 3.3 with respect to the different system rates.

G.826 Apportionment

The apportionment methodology divides the overall circuit into national and international portions, as shown in Figure 3.3.

The ITU-R has given the apportionment objectives for the international portion and national portions [13, 14].

International portion. The objectives for the international portion are given in Table 3.4.

The two factors under the operators control are the block allowance ratio B_R and the decision about whether one is operating as a terminal or intermediate country, which in turn determines the block allowance factor B_L. The distance allocation factor $F_L = 0.01 \times L_{km}/500$. The block allowance factor B_L depends on one's choice of country connection

$$\text{Intermediate countries:} \quad B_L = \begin{cases} B_R \times 0.02 \times L/L_{ref} & \text{for } L_{min} < L < L_{ref} \\ B_R \times 0.02 & \text{for } L > L_{ref} \end{cases}$$

$$\text{Terminating countries:} \quad B_L = \begin{cases} B_R \times 0.01 \times L/L_{ref} & \text{for } L_{min} < L < L_{ref} \\ B_R \times 0.01 & \text{for } L > L_{ref} \end{cases}$$

(3.e)

The reference length L_{REF} has a provisional value of 1000 km.

Table 3.3
G.826 Objectives

Rate (Mbit/s)	1.5–5	>5–15	>15–55	>55–160
Bits/block	2000–8000	2000–8000	4000–20000	15000–30000
ESR	0.04	0.05	0.075	0.16
SESR	0.002	0.002	0.002	0.002
BBER	3×10^{-4}	2×10^{-4}	2×10^{-4}	10^{-4}

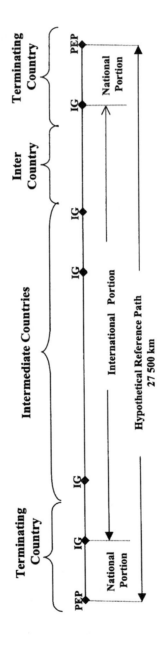

Figure 3.3 Apportionment of G.826 objective.

Table 3.4
International Portion of Radio Objectives

Rate (Mbit/s)	1.5–5	>5–15	>15–55	>55–160
ESR	$0.04(F_L + B_L)$	$0.05(F_L + B_L)$	$0.075(F_L + B_L)$	$0.16(F_L + B_L)$
SESR	$0.02(F_L + B_L)$	$0.02(F_L + B_L)$	$0.02(F_L + B_L)$	$0.02(F_L + B_L)$
BBER	$3 \times 10^{-4}(F_L + B_L)$	$2 \times 10^{-4}(F_L + B_L)$	$2 \times 10^{-4}(F_L + B_L)$	$2 \times 10^{-4}(F_L + B_L)$

National portion. In the national portion of the network the objectives are specified for long-haul, short-haul, and access networks. The SESR objectives are shown in Table 3.5.

Practical Advice

Planning engineers are often expected to design networks to meet G.821 or G.826 objectives, yet the apportionment assumptions or the range of parameters within those apportionment rules are seldom specified. Sensible assumptions thus need to be made. The actual conditions given in a hypothetical path will seldom be present in real networks unless the planning network is for a large PTT. The radio planner needs to consider the type of service being provided and what the quality of the alternative service is against which the new service would be compared. In the author's opinion, this is the key aspect to consider when choosing a standard to apply. If a backbone radio system is going to carry traffic that could also be carried over fiber, the highest quality standard should be chosen. If there is no alternative transmission medium and the service is noncritical, such as a voice service to a rural area, the lowest quality standard can be used. The type of network, on its own, is not a good guide to the quality level. For example, it has already been stated that a GSM network should be designed to a local-grade standard; yet if the transmission network is to carry other services, it is advisable to design the transmission network to a higher standard.

Table 3.5
National Portion of Radio Objectives for $A = A1 + 0.01 \times (L/500)$, where $1\% < A1 < 2\%$, and $7.5\% < B, C < 8.5\%$

SESR	1.5–5 Mbit/s	>5–15 Mbit/s	>15–55 Mbit/s	>55–160 Mbit/s
Long haul	$0.002 \times A$	$0.002 \times A$	$0.002 \times A$	$0.002 \times A$
Short haul	$0.002 \times B$	$0.002 \times B$	$0.002 \times B$	$0.002 \times B$
Access	$0.002 \times C$	$0.002 \times C$	$0.002 \times C$	$0.002 \times C$

References

[1]   ITU-R F.390-4, Geneva, 1982.

[2]   ITU-T G.821, Geneva, 1996.

[3]   ITU-R F.556-1, Geneva, 1986.

[4]   CCIR Report 445-3, Geneva, 1990.

[5]   ITU-R F.557-3, Geneva, 1992.

[6]   ITU-R F.695, Geneva, 1990.

[7]   ITU-R F.696-1, Geneva, 1991.

[8]   ITU-R F.757, Geneva, 1992.

[9]   ITU-R F.697-1 Annex 1, Geneva, 1991.

[10] ITU-R F.594-3, Geneva, 1991.

[11] ITU-R F.634-3, Geneva, 1994.

[12] ITU-R F.697-1, Geneva, 1991.

[13] ITU-R F.1092, Geneva, 1994.

[14] ITU-R F.1189, Geneva, 1995.

4

Radio Equipment Characteristics

4.1 Introduction

For the radio planner it is important to understand how radio systems work because the equipment characteristics dramatically affect the overall performance. Radio link performance standards are derived from ITU-T circuit-based standards that define limits for end-to-end circuits. An attempt will be made in this section to provide a basic understanding of what happens to the signal from one end-user to the other. A voice channel has been chosen to illustrate this; therefore, the concept of PCM has been included. The path of a voice circuit over a radio system is illustrated in Figure 4.1.

4.2 Configurations

Modern microwave radio equipment can roughly be divided into three categories: indoor, split unit, and all-outdoor. The primary multiplexer is usually external to the radio.

4.2.1 All Indoor

Traditional microwave equipment is housed in 19-in racks (21 in the United States) in a transmission equipment room. A coaxial or waveguide connection then transports the RF signal to the antenna mounted on a tower. The equipment is often of a modular construction for maintenance purposes. Different designs are normally required for different capacities and frequency bands. All indoor equipment is ideally suited to longhaul routes that require high-output

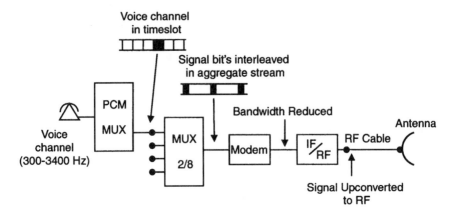

Figure 4.1 Transmission path of voice circuit.

powers and multifrequency branching arrangements. A typical layout is shown in Figure 4.2.

4.2.2 Split Unit (RF Outdoors)

Modern microwave equipment has moved away from the traditional approach consisting of large racks of equipment mounted indoors. It caters for access networks where space is limited and equipment commonality is preferred. Based on high-frequency (e.g., 23 GHz) link architecture that has the RF circuitry mounted outdoors to avoid the very high waveguide losses, equipment is now available at most frequencies and capacities in a split mount arrangement. In this arrangement the baseband and modem circuitry is mounted in an indoor unit, which is usually independent of frequency. This is connected to the outdoor unit that houses the RF circuitry via a low-cost coaxial cable. The cable carries the baseband or IF signal in addition to power and housekeeping signals. Phase-modulated systems require an IF signal for the up-and-down connection, whereas FSK systems can transport a baseband signal up and down the cable. The outdoor unit is often independent of capacity. The split-unit configuration is shown in Figure 4.3.

4.2.3 All Outdoors

With microwave radios being used to backhaul microcells in cellular networks, there is a requirement to be able to mount a radio link in a roadside cabinet. The antenna needs to be physically small and the radio link should not draw much power. Radio equipment operating at, for example, E1 or T1 line rates are already commercially available that can be mounted all outdoors. The base

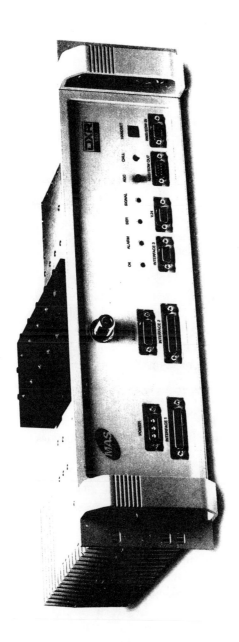

Figure 4.2 Typical all-indoor configuration.

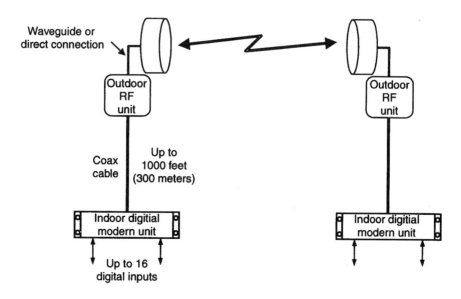

Figure 4.3 Typical split-unit configuration.

station equipment can be fed directly into the radio and can be co-located in the same enclosure. One obvious consequence of an all-outdoor radio is that, if there is a requirement to extend the E1 circuit to another location, a multicore cable will be required to carry the traffic, alarm, management, and power signals. An example of all-outdoor equipment is shown in Figure 4.4.

4.2.4 Basic Radio System Block Diagram

The various building blocks that make up a radio system are shown in Figure 4.5.

The end-user service (voice or data) is fed to the primary multiplexer where it is converted to a digital 64-kbit/s signal and multiplexed into an E1 (or T1) signal. This signal is then converted to the overall transmission capacity by a secondary multiplexer. An overhead is added to the transmission data rate to carry various services, and this aggregate proprietary radio baseband signal is then modulated and up-converted to the RF carrier frequency. The signal is then fed to the antenna for transmission. In the receive direction the signal is captured by the antenna and filtered via the branching unit to be fed to the receiver where it is down-converted to an IF signal and demodulated. The services are removed from the transmission data rate and the various signals demultiplexed back to the original form. The primary multiplexer converts the digital stream back to the original data or audio signal.

Figure 4.4 Typical all-outdoor radio.

4.3 Primary Multiplex

One would require infinite bandwidth to transmit a human voice over a transmission system without any distortion. The human voice does not, however, have an even distribution of energy. Most of its energy is distributed across the frequency spectrum from about 100 Hz to 6000 Hz. The maximum spectral density occurs at about 450 Hz for a male and 550 Hz for a female voice, on average. It has been found that by bandlimiting the voice signal to 300 Hz $< f <$ 3400 Hz, a high-quality voice signal can be transmitted. This is what is known in telephony as an audio channel. In digital systems our goal is to convert the analog voice signal into a digital signal. The process that is used is known as pulse code modulation (PCM). The PCM process has four components: sampling, quantizing, coding, and time multiplexing.

4.3.1 Sampling

The process of sampling is done by multiplying (or mixing) the analog signal with a periodic pulsed sampling signal. The process is illustrated by Figure

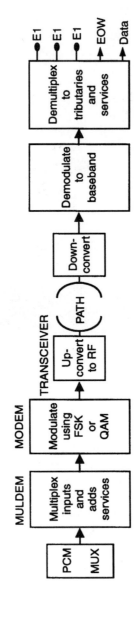

Figure 4.5 Block diagram of a microwave radio system.

4.6, which shows the original analog voice and pulsed sampling signals as well as the resultant sampled signal in both the time and frequency domains.

It is important to realize that the sampling process causes no distortion. The original signal is band limited and therefore, provided it is sampled fast enough, can be completely reproduced in the analog pulses. After sampling, the signal will be present as upper and lower sidebands around the harmonics of sampling signal. Using a lowpass filter, the original signal can be filtered out. The Nyquist sampling theorem says that no information content is destroyed in a band-limited signal as long as the sampling signal is at least twice the highest frequency component in the signal. If the sampling frequency is increased beyond this, the sidebands will just move further apart, making the original signal easier to filter out but not increasing the signal quality.

A sampling frequency of 8 KHz has been chosen by the ITU-T with a tolerance of 50 ppm. The highest frequency, assuming a perfect lowpass filter

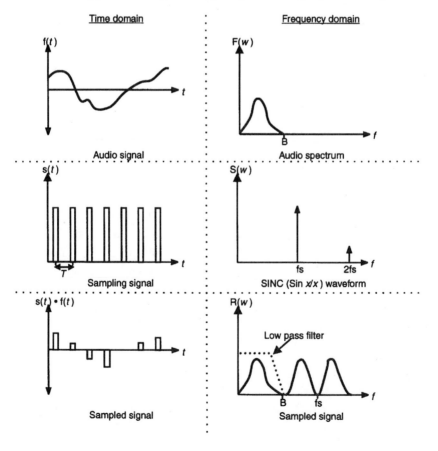

Figure 4.6 Sampled signal in time and frequency domains.

is thus 4 kHz (half of 8 kHz). Filters are not perfect; therefore, this frequency allows for a guard band of 600 Hz for filter rolloff. The filter that limits the audio signal to 3400 Hz is called an anti-aliasing filter.

4.3.2 Quantizing

Quantization is the process of obtaining a discrete value from an analog value. Recall that the sampling process converted a continuous-wave analog signal into a series of analog pulses. These pulses contain all the information present in the original band-limited signal. In the quantization process the values of the analog pulses are mapped into discrete levels. This process is required in order to have a limited number of sample values to code into a digital word. Eight to 16 levels are required for intelligible speech and 128 levels are required for high-quality speech. This is a one-way approximation and does cause distortion, as once the sample has been quantized, it is impossible to reproduce an exact replica of the original signal. The distortion caused by this process causes quantization noise. If uniform quantizing levels were chosen, a signal with a high value would have a better S/N ratio than one with a small value. In reality, the opposite is desired. High-amplitude pulses are easier to hear and, therefore, can tolerate a higher noise level. This problem is solved through companding.

4.3.3 Companding

Companding is an acronym for *comp*ressing and exp*anding*. A nonlinear algorithm is used whereby more quantized values are allocated to the smaller value samples, thus achieving a relatively constant error ratio for all samples. In Region 1 of the ITU (Europe and Africa), the A-law quantization curve is used to map the samples. This is a 13-segment characteristic whose characteristics are defined by the ITU [1]. In Region 2 (America), a 15-segment law called the μ-law is used. The positive half of the A-law companding characteristic is shown in Figure 4.7.

The A-law algorithm was implemented in the 1960s when circuitry could not achieve a logarithmic curve. A piecewise linear approximation was thus used. It can be seen from Figure 4.7 that half of the input voltage range is mapped into 16 quantized levels, the next quarter in terms of input voltage is also given 16 levels, the next eighth is given a further 16 levels, and so on. Small amplitude values are thus given much more quantized values, resulting in an improved S/N ratio. Using this approach, the S/N ratio of all samples, large and small, is made more even.

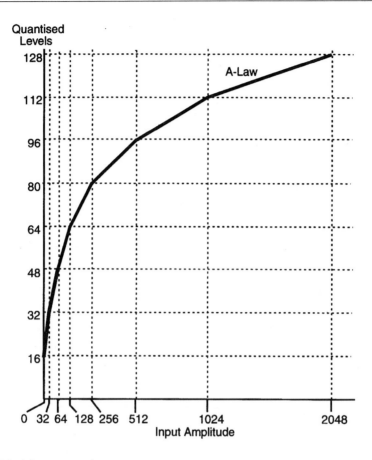

Figure 4.7 A-law compander curve.

4.3.4 Coding

The quantizing process using the A-law produces a total of 256 values (±128). This can be coded into binary form using eight bits ($256 = 2^8$) from the companding curve, as shown in Figure 4.8.

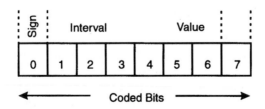

Figure 4.8 Coded bits from A-law compander curve.

The sign bit designates whether the sample has a positive or negative amplitude value. The next three bits designate in which of the eight intervals the value lies. The final four bits designate which sixteen values is closest to the actual sample. The coding process results in an eight-bit code byte that is a digital representation of an audio channel.

4.3.5 Time Multiplexing

The final process is to time multiplex the signals into a framed signal. The sampling is done at 8 kHz, that is, 8000 samples/sec for each audio channel. The sampling interval can be calculated from

$$T = 1/f$$
$$= 1/8000 \qquad\qquad (4.1)$$
$$= 125 \ \mu s$$

The duration of each pulse is 3.9 μs; therefore, it is possible to transmit samples from other audio channels in the time gap between the various samples. Before the second sample from the first channel is available for transmission, 32 samples (125 μs/3.9 μs) can be time interleaved. This is known as time division multiplexing.

Each sample consisting of eight bits occupies what is known as a timeslot within the frame. With a sampling speed of 8 kHz, an eight-bit timeslot thus has a transmission bit rate of 64 kbit/s. This is a fundamental rate within telecommunications systems and is known as E0 (Europe) or T0 (United States). A 32-timeslot frame has a transmission bit rate of 2048 kbit/s (2 Mbit/s). Usually only 30 timeslots are available for user channels because timeslot 0 is used for frame alignment and timeslot 16 for signaling. This 30-channel (or sometimes 31 channel) signal is known as E1. In the United States a frame rate based on the Bell system uses 24 64-kbit/s channels plus an extra framing bit to form a 1.544-Mbit/s signal. This is known as T1 or DS-1 (digital signal, level 1).

4.3.6 Primary Multiplex Equipment

A primary multiplexer used for voice circuits is often called a channel bank and converts 30 (or 24) voice channels into a framed E1 (or T1) circuit. With data becoming more prevalent in networks, a mixture of voice and data is often required to be multiplexed into the framed rate. Flexible multiplexers with a range of voice and data interfaces are thus available. Voice options include two wire, four wire, six wire (four W plus E&M signaling) for subscriber or exchange interfaces. Data interfaces include synchronous, asynchronous, ISDN, X.25, or ADPCM options. The more sophisticated multiplexers perform circuit grooming and allow cross-connect and management features.

4.4 Muldem (Secondary Multiplexing and Services)

The standard E1 (or T1) output from a primary multiplexer is not the only signal used in transmission networks. In some cases an external secondary multiplexer is required, for example, to create a 34-Mbit/s signal (E3) suitable to carry TV signals. The input to a radio system is usually one or more standard line rates such as E1, T1, E3, or STM-1. The radio system needs to transport these signals transparently to the other end of the link. In other words, it should not tamper with the signal in any way. The first thing the radio needs to do is to create a composite signal from the various inputs, which it can modulate and transmit to the other end. It needs to multiplex the various inputs and add any overhead that is required.

4.4.1 Multiplexing and Demultiplexing

In the very early digital radios, multiplexing was done externally to the radio. The requirement for transporting multiple E1s, however, led to the radio manufacturers including the secondary multiplexing function in the radio itself. A typical application is a 4- by 2-Mbit/s (4E1) radio where a 2- to 8-Mbit/s multiplexer is built into the radio. A composite 8-Mbit/s signal, which does not need to have a standard ITU interface, thus reducing cost and complexity, is available internally to the radio to be modulated and transmitted to the other end. In the opposite direction, the 8-Mbit/s signal is demultiplexed into the four E1 streams with a standard interface according to ITU G.703. The PDH secondary multiplexing function will be discussed in Chapter 9, where comparisons with SDH are made.

4.4.2 Overhead Channels

The radio system does have other signals that it needs to transmit. This could be for internal use of the radio's housekeeping functions, extra data channels for data or supervisory systems, an engineering order wire (EOW), and forward error correction (FEC), as discussed below. To do this, a complementary radio overhead channel is usually added to the signal to produce an aggregate rate that exceeds the ITU line rate. This signal is a proprietary signal that could be different for each manufacturer. It obviously adds bandwidth to the baseband signal and therefore will be kept as low as practically possible so that the aggregate signal fits into the required channel bandwidth. An example of a radio complementary overhead forming a proprietary aggregate signal is shown in Figure 4.9.

4.4.2.1 Data Channels and Supervisory

Most radio systems can carry a limited number of data channels on the radio overhead capacity. Various low-speed or high-speed data channels, either synchronous or asynchronous, are catered to. The data is usually carried on one or two 64-kbit overhead channels that determine which options can be used simultaneously.

4.4.2.2 Engineering Order Wire

An EOW is available on most equipment to enable maintenance staff to be able to communicate over the radio without using the multiplexer equipment. A standard headset, usually dual-tone multifrequency (DTMF), is usually used with an analog two-wire interface. The overhead channel is usually carried over a 64-kbit/s channel for a 300- to 3400-Hz audio signal. If high-quality transmission is not required, more than one order wire channel can be provided over a single 64-kbit/s channel using a reduced-rate (usually 300- to 2400-Hz) audio bandwidth. One needs to be careful not to cascade too many EOWs on the route because the quantizing noise from the analog-to-digital conversion process can make the channel noisy. Signaling can be provided using E&M for six-wire circuits and inband DTMF for two-wire and four-wire circuits.

4.4.2.3 Forward Error Correction

FEC is standard on most modern microwave radio systems to meet the high-quality background error rate objectives. FEC is a technique of detecting and correcting errors that may occur over the link. FEC, based on block coding, uses extra bits carried in the radio overhead to do the error detection. Error correction using some form of trellis coding, where the coding is done in the modulation phase without adding bit overhead, is not discussed here. Simple

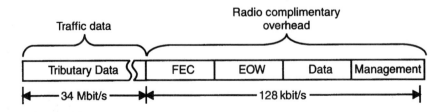

Figure 4.9 Radio overhead using bit insertion.

FEC based on adding overhead bits is done by performing a mathematical algorithm on the bits and transmitting this code to the opposite end. If a different code is received, an error has occurred. Only a limited number of bit combinations are allowed by the coding technique; thus, not only can errors be detected but a limited number of errors can be corrected. A Reed-Solomon code with 20 correction bytes out of 224, for example, will correct up to 10 bytes in each block. FEC does not provide much improvement under fading conditions but has excellent performance against dribble errors, as shown by Figure 4.10.

Modern radio systems can achieve residual error rates comparable to fiber systems, that is, better than 10^{-12}.

4.4.2.4 Wayside Channels

In medium- to high-capacity applications, such as 34-Mbit/s systems, one often wants to avoid having to install an expensive 2 to 34 multiplexer at a site if access to only one E1 channel is required. For this reason, radio manufacturers often offer a single user channel carried on the radio overhead that can be transported to a nodal site and then incorporated into the main traffic.

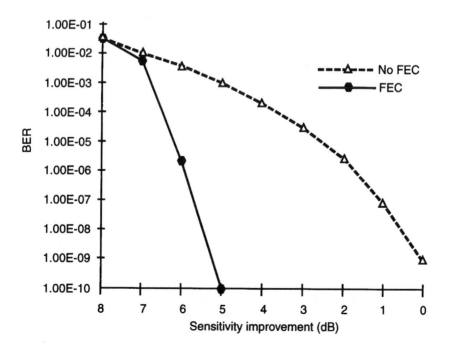

Figure 4.10 Typical FEC curve showing system improvement.

4.4.3 Baseband Filtering

Baseband filtering is done to limit the bandwidth of the signal. The shaping of the baseband signal is very important. Infinite bandwidth would be required to ensure that the input pulses were not rounded off in any way, which is obviously not possible or desirable. A practical filter that results in zero crossing points at the Nyquist frequency is a raised cosine filter [2]. The bandwidth of a multilevel modulation signal with baseband shaping can be calculated from

$$BW = [\text{baseband bit rate}/\log_2 M] \cdot (1 + \alpha) \qquad (4.2)$$

where α is the filter rolloff factor and M is the M-ary modulation value (e.g., 16-QAM, $M = 16$). As implied earlier it is important that the filtering of the signal does not result in intersymbol interference from the leading and trailing tails of the signal. A Nyquist filter with a rolloff factor of 0.5 is usually used that ensures the tails of the adjacent pulses are at zero during demodulation. This Nyquist pulse shaping with the tails crossing zero at the point of sampling is illustrated in Figure 4.11.

4.4.4 Basic Muldem Block Diagram

The various building blocks that make up the muldem section of a typical radio system are shown in Figure 4.12.

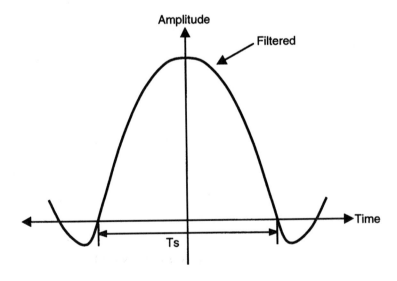

Figure 4.11 A typical baseband pulse with Nyquist shaping.

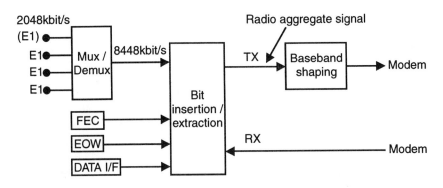

Figure 4.12 A block diagram of the Muldem section of a radio system.

4.5 Modem

A modem is a word that is shortened from *mo*dulator–*dem*odulator. The baseband signal has to be transported over a radio frequency carrier, and it is done by modulating the baseband signal onto an IF or RF carrier.

4.5.1 Modulators

4.5.1.1 Types of Modulation

Two main types of modulation exist for digital radio systems, namely, direct modulation or indirect modulation.

Direct modulation is when no IF carrier exists. The baseband signal is applied directly to the modulator, thus reducing cost and complexity. Indirect modulation involves first converting the baseband signal to an IF and then converting this to an RF frequency. There are three main types of digital modulation: amplitude, phase, or frequency modulation. Since it is a digital signal, this modulation switches the signal between two states. In amplitude modulation, on-off keying (OOK) is used when the amplitude value is switched between zero and some predetermined amplitude; in phase modulation (PSK) the phase is shifted by 180 degrees; and in frequency modulation the frequencies are shifted between two frequency values. The two most commonly used modulation methods for microwave radio equipment are based on multilevel FSK and *n*-QAM, which are based on a combination of the methods discussed previously. These schemes use multisymbol modulation to reduce the bandwidth requirements. Multisymbol modulation schemes, while requiring a higher S/N ratio to operate, halve the bandwidth requirements for each level used.

4.5.1.2 FSK

FSK is a cost-effective and robust modulation scheme. It is not sensitive to amplitude and phase variations (noise and jitter) and hence does not require transmitter backoff. Higher transmit output powers are thus possible. The signal can be directly modulated onto the RF carrier without the need for an IF frequency, thus simplifying the circuitry and reducing cost. Noncoherent (nonphase synchronous) receivers can be utilized. Frequency modulated detectors have much simpler circuitry because they are far less affected by amplitude and phase variations than coherent schemes. A low-cost modem can thus be provided with adequate system gain.

4.5.1.3 QAM

Coherent demodulators provide improved receiver thresholds; therefore, to maximize system gain, phase modulation is often chosen despite the added cost and complexity. For bandwidth-efficient medium- to high-capacity systems, QAM is the preferred modulation. Let us start by considering a basic bi-phase shift keying (B-PSK) system. A carrier signal is switched in phase by 180 degrees to represent the binary string of 0s and 1s. If one plots this on a phase diagram it would appear as shown in Figure 4.13.

In order to halve the bandwidth with multisymbol modulation, a second B-PSK modulator could be employed operating in quadraphase to the first. If the incoming binary stream was divided in two by sending the alternate bits to the pair of B-PSK modulators, four different phase alternatives would exist, as shown in Figure 4.14.

Note that in QPSK the phase vectors each have the same amplitude; it is only the phase that is different. In practice, this is sometimes called differential phase shift keying (DPSK) since it is not the absolute phase value that is used but the difference between two phase states. Now consider 16 QAM. In this case the incoming bit stream is divided into four paths with each phase modulator handling four bits at a time. An example of what the constellation diagram could look like including the four-bit values is shown in Figure 4.15.

The phase vectors not only differ in phase but also in amplitude, thus halving the bandwidth requirement compared to QPSK. One should note, however, the distance that the phase vector can move before being decoded as a different bit sequence is now much less than with QPSK. In other words, the minimum S/N ratio required at the demodulator is greater with higher

Figure 4.13 Phase diagram showing PSK modulation.

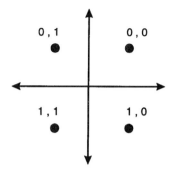

Figure 4.14 QPSK constellation diagram and circuit block diagram.

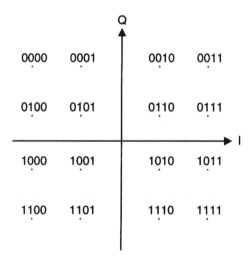

Figure 4.15 16 QAM constellation diagram.

levels of modulation. The modulation decision is thus a tradeoff between narrower bandwidth and performance. High-capacity SDH systems use 128 QAM, which allows a 155-Mbit/s signal to fit within a channel bandwidth of 28 MHz.

4.5.2 Demodulators

4.5.2.1 Types of Demodulators

Two main types of demodulators are used to detect digital signals: envelope detectors and coherent demodulators. Envelope detectors use simple diode detectors to extract the envelope of the signal. For phase-based systems such as PSK or QAM, there is no envelope variation; therefore, coherent (synchronous) demodulation is required. In this method the incoming modulated carrier signal is mixed with an exact replica (phase and frequency) of the IF carrier. A lowpass filter is then used to recover the original baseband signal. The required replica carrier is generated using a Costas loop, which uses a phase lock loop (PLL) to stabilize the carrier frequency extracted from the incoming RF signal, down-converted to IF. In addition to this recovered IF signal a baseband clock signal is recovered for the demodulation process. This type of demodulator is more expensive due to the complexity of achieving phase synchronism, but they do result in improved receiver thresholds.

4.5.2.2 Adaptive Equalization

In order to overcome the effects of dispersive fading caused by multipath conditions, longhaul radios will often use equalizers to reduce the fading effects. Static frequency-based equalizers can be used at IF frequencies to equalize the frequency response. This is usually done using simple slope and bump circuits. For example, if three notch filters are used to detect the amplitude level across the receiver bandwidth, a slope or notch can be detected. By generating the opposite slope or bump, the response can be equalized. A more powerful technique that can equalize the phase response is done in the time domain. This is called transversal adaptive equalization (TAE). The basic concept is to use a series of shift registers as a delay line. The distortion from the delayed signal can thus be detected, and by adding the correct delay factors and tapping the signal back on itself, the distortion can be equalized. This needs to be done for both minimum phase and nonminimum phase conditions; therefore, feedback and feedforward taps are required. In the past a combination of analog and digital delay lines were used for the positive and negative delays, respectively, causing an uneven signature curve response for the two conditions. Since digital delays lines are now used, the signature curve is usually the same for both minimum and nonminimum phase conditions. The greater the number the taps on the equalizers, the better the performance.

4.5.3   Basic Modem Block Diagram

The various building blocks that make up the modem section of a typical radio system are shown in Figure 4.16.

4.6   Transceivers

The RF section that comprises the transmitter and receiver modules is known as the transceiver.

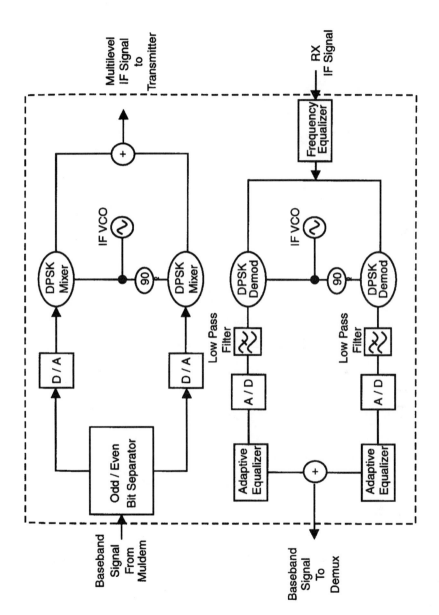

Figure 4.16 A block diagram of the modem section of a radio system.

4.6.1   Transmitters

Once the incoming signals are multiplexed and combined with the overhead channels, the baseband signal is modulated as discussed previously. This signal is then up-converted to the RF carrier frequency and amplified with a power amplifier. Modern transceivers are synthesized, meaning that a reference oscillator is used to derive the RF frequencies using a local oscillator that is voltage controlled (VCO). Using the synthesized VCO, transceiver frequencies can be selected by software across a broad range. The power amplifier is designed to be as linear as possible, however, it will still introduce some signal distortion. To keep this to a minimum the signal is often predistorted prior to amplification, thus introducing complementary distortion products that are canceled in the power amplifier. Linearity is so important that, although power amplifiers could amplify the signal up to the saturation level, a transmitter "back-off" is purposely applied that improves linearity and hence receiver threshold. The transmitter also usually has an automatic gain control (AGC) circuit to keep the output power constant as temperature variations occur.

4.6.2   Receivers

In the receive direction, the RF modulated carrier is down-converted to an IF frequency before demodulation. This is done by mixing the RF frequency with the synthesized VCO local oscillator frequency. An AGC circuit ensures that the IF output is kept constant as the RF signal level varies. This AGC signal is thus often used to measure the strength of the receive signal.

4.6.3   Basic Transceiver Block Diagram

The various building blocks that make up the transceiver section of a typical radio system are shown in Figure 4.17.

4.7   Branching

The branching unit is a generic term to describe the circuitry that interfaces the antenna to the transceiver. It includes filters, combiners, and isolators.

4.7.1   Duplexer

The same antenna is used for both transmit and receive frequencies. The branching unit filters the signals and combines the two signals onto one antenna. Filtering the transmit signal is done to ensure that the spectrum transmitted

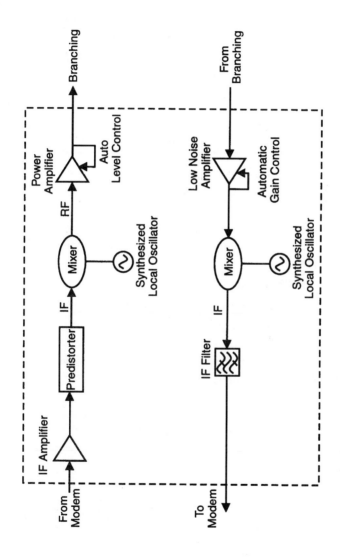

Figure 4.17 A block diagram of the transceiver section of a radio system.

does not cause interference on the adjacent channels. Various standards limit the allowable transmit spectrum for each band of operation. In the receive direction, the signal is filtered to eliminate any spurious signals being transferred to the receiver circuitry for demodulation and to limit the thermal noise, which is proportional to receiver bandwidth. RF filtering at this level is fairly basic because in modern radio systems the RF spectrum is limited by baseband filtering, thus shaping the signal into the required spectrum occupancy.

Combining transmit and receive signals onto one antenna is achieved by a device called a circulator. The combination of circulator and filter is often called a duplexer or diplexer. A circulator transfers the signal with very low loss to the wanted port while giving a high isolation to the unwanted signal on the other port. The transmit signal therefore is transferred to the antenna with low loss and very low leakage into the receiver with the same situation in the receive direction, as shown in Figure 4.18.

It is very important for the radio planner to understand branching losses and to include them in the design calculations. One needs to carefully check specification sheets to determine if the output power, for example, includes the branching loss or not. It will not be possible to accurately predict the expected receive level if branching losses have not been included.

4.7.2 Hotstandby Branching

In a hotstandby arrangement only one frequency pair is used for the two radio systems. It is thus not possible to transmit both systems simultaneously. A transmit switch is required to transmit one or the other transmit signals. Both transmitters actually transmit a signal, but only one is switched onto the antenna. The other signal is transmitted to a dummy load. This could reduce the transmit signal by up to 0.5 dB. In the receive direction, the signal is split into two paths and both signals are demodulated with the best signal being selected. Splitting the signal into two paths means that each signal is reduced by 3 dB, however, in practice this loss is typically 3.5 dB to 4 dB.

A diagram of a hotstandby arrangement is shown in Figure 4.19.

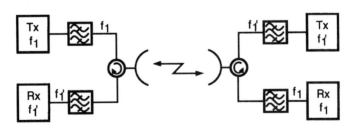

Figure 4.18 Duplexer in branching unit for a 1 + 0 configuration.

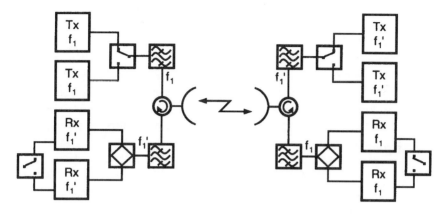

Figure 4.19 Block diagram of HSB branching.

4.7.3 Frequency Diversity Branching

In frequency diversity both transmitters are transmitted simultaneously and each signal is fed to its respective receiver without a transmit switch or a receiver hybrid. The losses are thus significantly less than with a hotstandby arrangement. Circulator losses and filter losses are typically only 0.1 dB each. The branching diagram is shown in Figure 4.20.

4.7.4 Space Diversity Branching

With space diversity only one frequency pair is used. Only one transmitter needs to be connected, however, in order to have equipment protection as well

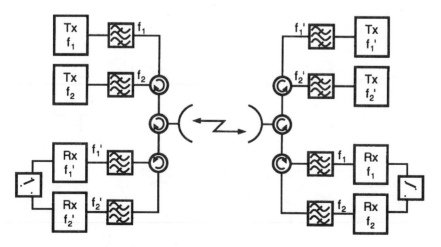

Figure 4.20 Block diagram of FD branching.

as path protection, the transmit branching is often the same as the hotstandby arrangement. Usually the top antenna is used for the transmit path. In the receive direction two antennas are used that are each fed to their respective receivers. One transmit and two receive antennas are required in each direction. Thus, a total of four antennas are required. The branching arrangement for a typical space diversity system with baseband switching is shown in Figure 4.21.

4.7.5 Hybrid Diversity Branching

For extra performance on very long or difficult paths, frequency diversity and space diversity can be combined. This is called hybrid diversity. In a 1+1 frequency diversity system this can be very cost effective because only three antennas are required to give full space and frequency diversity improvement in both directions. This is done by transmitting the transmit signal from the second frequency diversity path on the lower antenna at the one end. The arrangement is shown in Figure 4.22.

The best diversity improvement can be obtained using four antennas and four receivers, since there are then three separate paths (with corresponding improvement factors) that can be considered. This is shown in Figure 4.23.

4.8 Equipment Characteristics

Link planners need to be aware of the characteristics of radio equipment in order to specify the correct equipment and use the correct parameters in the radio link design. The most important characteristics that are normally included on a data sheet are discussed in the following sections.

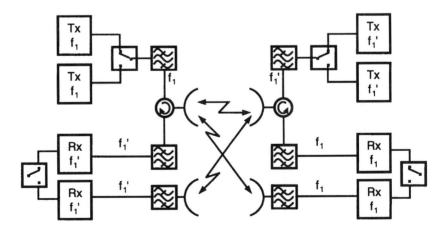

Figure 4.21 Block diagram of SD branching.

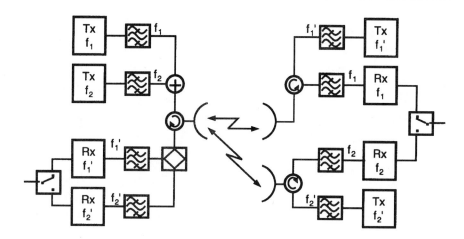

Figure 4.22 Block diagram of hybrid diversity with three antennas.

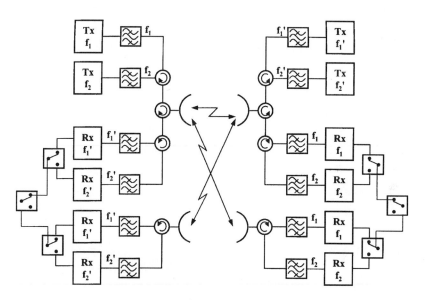

Figure 4.23 Block diagram of space diversity with four receivers.

4.8.1 RF Details

4.8.1.1 Frequency Range

Radio equipment is designed to operate over a certain frequency range. Nonsynthesized equipment will be tuned to the actual channel that is being used before delivery to site. Synthesized equipment can be software tuned to the channel frequency on site; however, even though it will operate over a large frequency range, this does not always cover the entire frequency band, so more than one transceiver may be required. Different transceivers are also usually required for the high- and low-band ends. The equipment transceiver range should thus be checked against the frequency plan that is being used. One should then determine what has to be done to tune the radio to the specific frequency channel. This will also involve different branching requirements.

4.8.1.2 Tx/Rx Separation

The minimum spacing allowable by the radio will be specified. This is a function of the RF filtering and branching isolation. The radio planner needs to check the equipment specification against the frequency plan being used.

4.8.1.3 Channel Spacing

One needs to check that the channel spacing required is supported by the equipment. The filtering and modulation technique will determine the channel spacing. Channel filters, which form part of the branching, are often required at lower frequencies (e.g., 7 GHz) to meet the strict bandwidth limits set by the ITU.

4.8.1.4 Frequency Stability

The stability of the RF carrier is normally specified in parts per million (ppm). One ppm corresponds to 1 Hz in a megahertz or 1 kHz in a gigahertz. If the stability of a 7-GHz carrier is given as 3 ppm, the offset allowed is 21 KHz.

4.8.2 Transmitter Characteristics

4.8.2.1 Transmit Output Power

The transmit output power is usually specified either at the transmit output module or at the antenna flange in dBm. In the latter case, the transmit branching losses are already included. One should check whether typical or guaranteed figures are specified. Typical figures tend to be 3 dB to 4 dB better than guaranteed.

4.8.2.2 Transmit Power Control

The transmit output power can often be attenuated using software settings in the radio. An adaptive transmit power control called automatic transmit power control (ATPC) is used to improve frequency interference by attenuating the transmit power under nonfaded conditions and then boosting the power during fading. This is done by monitoring the receive level and feeding this information back to the transmitter. If no fading occurs, the transmit power is attenuated, thus reducing the EIRP. During fading this attenuation is removed, thus restoring the full design fade margin included to overcome fading effects.

4.8.2.3 Output Spectrum and Spurious Emissions

To reduce interference into other systems, the spurious emissions from a transmitter need to be reduced by adequate shaping and filtering. Transmit output masks and spurious emission limits relative to the carrier frequency are specified in equipment standards.

4.8.3 Receiver Characteristics

4.8.3.1 Receiver Threshold 10^{-6} and 10^{-3}

The receiver threshold is a critical parameter to obtain since this is one of the main parameters used to determine the fade margin. Strictly speaking, it is a 10^{-3} value that is used for the fade margin since outages are based on SES. Users often prefer the 10^{-6} value nowadays as a minimum quality level for data. One should use the guaranteed threshold values in the calculations. One must be clear on whether the quoted values are relative to the antenna flange or specified at the input to the receiver. Receiver threshold values are quoted in dBm. They will always be a negative value, typically around -70 dBm to -90 dBm.

4.8.3.2 Maximum Receive Level

For short hops one needs to be careful not to exceed the maximum input receive level. If the signal level is too strong, errors can occur due to saturation of the receiver circuitry. If the levels are extreme, irreversible damage can occur. Equipment manufacturers will specify the maximum overload receive level. Maximum levels are quoted in dBm, typically around -15 dBm.

4.8.3.3 Dispersive Fade Margin

The dispersive fade margins (DFM) are usually quoted for 10^{-6} and 10^{-3}. As with receiver threshold values, the 10^{-3} value is the correct one to use for the fade margin. Adaptive equalizers dramatically improve DFM values. The DFM value for equipment should typically be 10 dB better than the flat fade margin required. DFM values are quoted in decibels and vary from around 35 dB (without equalizers) to better than 70 dB.

4.8.4 C/I Ratio

Frequency planning requires some equipment parameters for the interference calculations. The minimum carrier-to-interference ratio (C/I) that the demodulator can tolerate is important, as is the net filter discrimination (NFD). This is covered in detail in Chapter 7. Manufacturers will normally supply curves or a table of values for these two parameters. In digital systems, interference at the threshold is more critical than under unfaded conditions; therefore, the threshold-to-interference (T/I) values are required.

4.8.5 Digital Interfaces

It is important to specify the radio equipment baseband interface required because different standards exist. The normal radio interface complies with ITU-T G.703 and can either be an unbalanced 75-Ω coaxial connection or a balanced 120-Ω twisted-pair cable connection. This often needs to be specified before delivery of the equipment; however, in some equipment both options are supported and are software selectable. High-capacity radios such as STM-1 have an optical connection to be interfaced to an ADM. An electrical option is often included as well.

4.8.6 Management and Alarm Interface

Modern equipment is software configurable using a PC. Usually a link can be set up, configured, and monitored without having to do any physical adjustments on the equipment. Another recent advancement is the use of a standard Web browser to access the radio link using an Ethernet connection. By providing each radio terminal with an IP address the radios can be accessed over the internet with the display appearing in HTML format. This allows one to access any element in the network via a remote PC. Security is obviously required to ensure that only authorized personnel have access to this information. A connection to the network management system is also required. Nowadays this is usually an Ethernet connection.

Most radios also have various alarm inputs and outputs. Inputs are required to carry the alarms from colocated external equipment over the radio system. This may also be station alarms such as door alarm or tower light alarm. Relay outputs are also sometimes provided for controls. For example, one may want to set off an audio or visual alarm at a station. These management and alarm interfaces are usually available from the front panel of the equipment in DB-type or Ethernet type (10-Base T) connectors.

4.9 Power Details

4.9.1 Input Voltage Range

Most microwave telecommunications equipment runs off 48-V DC; however, traditional low-capacity radios used 24V and therefore many sites still have

24-V power supplies. Some radio equipment have an extended input range that accommodates 24-V or 48-V supplies of both polarity, however, an external power converter may be required for equipment that does not cover this range. Equipment installed in urban areas also often does not have a DC power supply; therefore, inverters may be required for the equipment to run off a main supply. A small battery back-up should be included to overcome power cuts.

4.9.2 Power Consumption

In order to work out the requirements for the station's power supply and battery capacity one needs to add up the total power consumption of all the equipment. The power consumption figures of the radio terminals need to be considered. These figures are quoted in Watts.

4.10 Environmental Considerations

It is becoming increasingly important to comply with environmental specifications. In Europe it is mandatory that equipment complies to the strict electromagnetic compatibility (EMC) standards [3–5]. In addition, limits are set on aspects such as operational temperature range, ingress protection (water, humidity, dust), shock and vibration, and transportation and storage [6].

4.11 Equipment Type Approvals

In many countries a government-appointed telecommunications regulator exists who will often insist that equipment be type approved before being installed into a network. This usually involves proving conformance to relevant international equipment standards with particular emphasis on aspects such environmental aspects, EMC, and transmit output spectrum. The radio planner should ensure that equipment being used is approved for use in those countries that demand it.

References

[1]   ITU-T Recommendation G.711, Geneva.

[2]   Schwartz, M., *Information Transmission Modulation and Noise*, New York: McGraw-Hill, 1981.

[3]   European Directive 89/336/EEC.

[4]   European ETS 300 385 specification.

[5]   American Bellcore GR-1089-CORE specification.

[6]   European ETS 300 019.

5

Microwave Propagation

Link design is mainly about accurately predicting the outage period that a radio link will suffer and ensuring that it does not exceed the quality objectives discussed in Chapter 3. Most outages occur as a result of atmospheric effects; therefore, it is necessary for the radio planner to have a thorough understanding of microwave propagation.

5.1 Atmospheric Effects on Propagation

Many articles written about radio links imply that the beam is a pencil-thin ray that travels in a straight line between two antennas. In reality, it is an electromagnetic wavefront that is infinitely wide even with high-gain microwave antennas. The path that the wavefront travels is dependent on the density of the troposphere—the lower portion of the atmosphere—that it encounters. In a vacuum the density that the wavefront would encounter is uniform. In a so-called standard atmosphere the average density decreases with altitude. The upper portion of the wavefront thus travels faster than the lower portion that is traversing the denser medium. Since the direction of propagation of an electromagnetic wavefront is always perpendicular to the plane of constant phase, the beam bends downward. This is called refraction.

5.1.1 Refractive Index

The refractive index (n) is the ratio of the speed of an electromagnetic wave traveling in a vacuum relative to the speed it would travel in a finite medium as expressed by

$$n = c_0/c \tag{5.1}$$

where c_0 is the speed of light and c is the speed of microwave beam in the finite medium.

5.1.2 Radio Refractivity

The refractive index is always greater than unity, but for a radio wave traveling in air it is only a small fraction greater than unity. For example, the average ground refractive index is 1.000315, which is an inconvenient number, so a radio refractivity (N) has been defined

$$N = (n - 1) \times 10^6 \tag{5.2}$$

Substituting the value for the ground refractive index ($n = 1.000315$) into (5.2) yields a value for N of 315 N-units.

The radio refractivity for links below 100 GHz is defined as [1]

$$N = 77.6 \ P/T + 3.732 \times 10^5 \ e/T^2 \tag{5.3}$$

where P is the atmospheric pressure in mbars, T is the absolute temperature in Kelvins, and e is the partial pressure due to water vapor in mbars.

The value of N varies with altitude since pressure, temperature, and humidity all vary with height. Pressure and humidity normally decrease exponentially with height. Temperature normally decreases linearly with altitude of approximately $-6°$/km. Humidity and temperature do, however, change under certain conditions, resulting in variations of radio refractivity.

In general, the atmosphere displays an exponential decrease of N with height. For an average atmosphere the refractivity can be written as

$$N(h) = N_0 \exp(-h/h_0) \tag{5.4}$$

where $N_0 = 315$ N-units (average refractivity value extrapolated to sea level) and $h_0 = 7.35$ km.

5.1.3 Refractivity Gradient

As a radio link designer one is not so much interested in the absolute level of refractivity as in the change in the value over the microwave front. It is thus the gradient of refractivity that is of interest. Although the atmospheric gradient is exponential with altitude, in the lowest few hundred metres—which is where the axis of the microwave beam is traveling—it can be approximated as a linear gradient. The refractivity gradient is thus defined as

$$G = dN/dh \qquad (5.5)$$

Under well-mixed atmospheric conditions this value is a constant. Experimental results from a hop in Trappes (France) [2] show that the median value of the refractivity gradient for a typical hop length is −39 *N*-units/km. Refractivity gradients change with time, leading to anomalous propagation conditions, which are discussed in greater detail later in this chapter.

5.1.4 Effective Earth Radius

Due to refraction of the signal, the radio wave does not travel in a straight line. The bending of the ray is dependent on the point gradient of refractivity that the ray experiences at each point along the path. If one averages these point gradients over the path, it can be assumed that the ray follows a curved trajectory. The ray can thus be considered to travel in an arc with radius *r*. This radius is inversely proportional to the average refractive index gradient over the path, hence allowing the following approximation

$$1/r = dn/dh \qquad (5.6)$$

Just as the radio ray is not a straight line, the Earth's surface over which it travels is not flat. Even if traveling over a flat surface such as the sea, the curvature of the Earth needs to be taken into account. The Earth is not round but an oblate spheroid, however to simplify matters, the surface of the Earth can be approximated to be an arc with an average radius of 6,371 km.

We now have a situation where the clearance of the radio beam over the Earth's surface is dependent on the relative distance between two curves. An analysis of the clearance is made easier if one of the curves is straight and the other is given extra curvature to compensate. It is convenient to imagine that the radio ray travels in a straight line relative to an effective Earth radius, which has been adjusted by the refractivity gradient. This radius is the real Earth radius multiplied by an effective Earth radius factor "*k*" that is dependent on the refractivity gradient. This is commonly referred to as the "*k*" factor and should not be confused with the geoclimatic factor "*K*."

It is essential to understand that whenever one uses *k*-factor analysis, one is no longer dealing with the real scenario. One curve is fictitiously made straight and the other fictitiously given extra curvature. A *k*-factor analysis should be used to determine relative clearance and not to predict beam curvature in terms of angle-of-arrival into an antenna. For example, when the *k-factor* goes below unity, the actual beam bends upward whereas, using ray-tracing techniques, the beam would appear to bend downward.

Given this, what value do you assume the *k*-factor would be when the signal travels parallel to the Earth's surface? Intuitively one may feel the answer is unity ($k = 1$). However, because one of the curves is made straight, the *k*-factor must be infinity to compensate. A *k*-factor of unity means one of the curves is already straight; hence, it would refer to a radio signal traveling in a straight trajectory over a curved Earth equal to the real radius of the Earth.

The *k*-factor is related to the refractive index gradient [3]

$$k = 1/(1 + a \, dn/dh) \tag{5.7}$$

where *a* is the real radius of the Earth (6371 km).

If we rearrange the radio refractivity *N* defined in (5.2) we get

$$n = N \times 10^{-6} + 1 \tag{5.8}$$

To obtain the gradient we need to differentiate (5.8) with respect to height. Hence

$$dn/dh = 10^{-6} \, dN/dh = 10^{-6} \, G \tag{5.9}$$

Substituting this into (5.7) yields

$$k = 1/(1 + 0.006371 \, G) \tag{5.10}$$
$$= 157/(157 + G)$$

Using (5.10) one can relate some commonly referred to values of *k* to *G*, as shown in Table 5.1.

5.1.5 Anomalous Propagation

The refractivity gradients, which cause the bending of the radio beam, also change with time. Experimental measures have shown that the refractivity gradient can change from being positive to steeply negative. The extreme values only persist for small percentages of time. Typical field results are shown in Figure 5.1.

It can be seen from Figure 5.1 that extremely negative and positive values are only present for small percentages of the time. The median value (50%)

Table 5.1
Comparison of *k*-factor Versus Refractivity Gradient

k-factor	Radio Refractivity Gradient *G*
k = 1	*G* = 0
k = 4/3	*G* = −39
k = ∞ (infinity)	*G* = −157
k < 1	*G* > 0

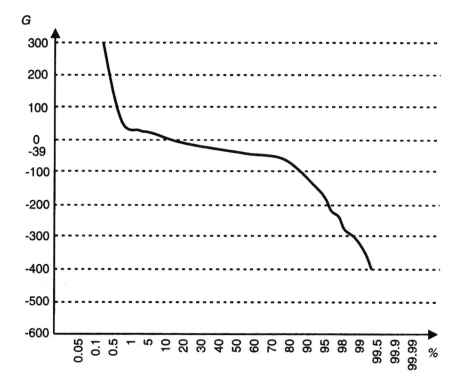

Figure 5.1 Field results showing probability distribution of refractivity gradients.

in this case is −39 *N*-units/km, which corresponds to a *k*-factor of 4/3. It can also be seen that the negative values are more extreme than the positive values. When the gradient of refractivity displays average characteristics, it is called standard refraction. This corresponds to the commonly quoted value of *G* = −39 *N*-units/km (or *k* = 4/3). When the gradient becomes positive it is known as subrefraction and can cause diffraction loss. When the gradient becomes more negative than $G = -100$ N-units/km, it is called super-refractive and results in multipath fading. When the gradient becomes more negative than $G = -157$, ducting conditions occur resulting in severe multipath, beam spreading, and even blackout conditions. The ITU [4] provides a series of curves that detail the percentage of time the refractivity gradient (P_L) is less than -100 N-unit/km. This gives an indication of the probability of ducting being a problem. These values are used in ITU Method 530 discussed in Chapter 8. The bending of radio waves caused by different refractivity gradients is shown in Figure 5.2.

5.1.6 Physical Atmospheric Conditions

In this section we discuss the physical reasons why these atmospheric layers display abnormal refractivity gradients. Recall from (5.3) that the radio refractivity (N) can be expressed in terms of pressure (P), temperature (T), and humidity (e). The effects will be discussed in terms of positive and negative refractivity gradients, respectively.

5.1.6.1 Positive Gradients

Considering (5.3) it can be seen that in order to have a positive refractivity gradient one requires either a strong negative temperature gradient or a positive humidity gradient or both. The various pressure, temperature, and humidity gradients causing positive refractivity gradients are compared with normal gradients in Figure 5.3.

The advance of cool moist air (advection) can cause positive gradients over hot, dry ground especially in coastal regions. This results in a steep increase in humidity above the ground surface. Another cause is the lifting of a warm air mass by cool dry air during frontal weather processes (storms). Autoconvection, which describes the convection of heat from an extremely hot surface, is another possible cause. In this case a steep negative temperature gradient is present in the hot dry air above the Earth's surface and bounded by the cooler air mass displaying normal refractivity. Conduction due to solar heating has the same effect.

5.1.6.2 Negative Gradients

Once again considering (5.3) one can see that negative refractivity gradients require positive temperature gradients (temperature inversion) and/or strong negative humidity gradients (hydrolapse). The various pressure, temperature, and humidity gradients, which result in abnormal negative refractivity gradients, are compared with normal gradients in Figure 5.4.

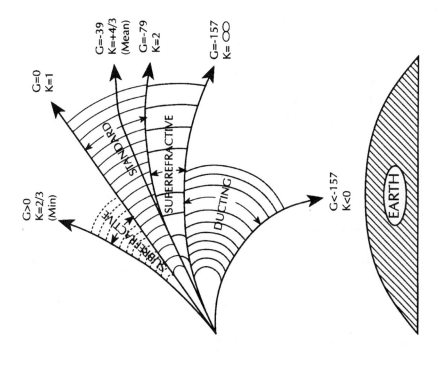

Figure 5.2 Ray bending caused by refraction.

Normal Propagation

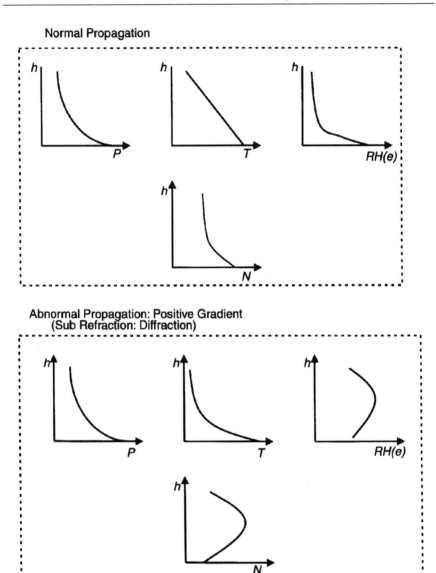

Figure 5.3 Positive refractivity gradients.

Abnormal negative gradients (< −100 N/km) can result due to various atmospheric conditions such as subsistence, advection or surface heating, and radiative cooling. These processes can lead to extremely steep negative refractivity gradients in excess of −157 N/km and result in the formation of ducts.

The first process that can lead to ducting is evaporation. A shallow surface-based duct can form above wet surfaces such as moist ground or the sea due

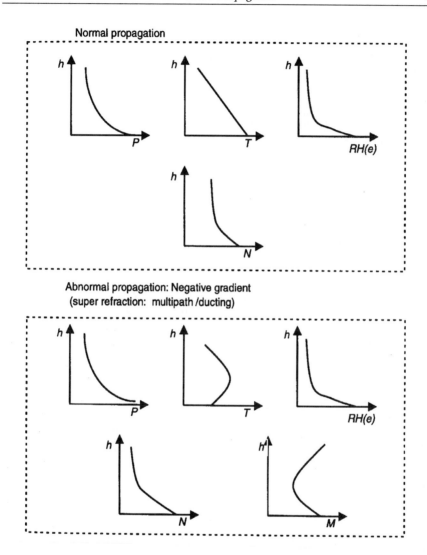

Figure 5.4 Negative refractivity gradients.

to the steep negative humidity gradient. In some areas, a semipermanent duct can form above the ocean, with depths up to 20m. The depth varies according to altitude (the lower altitudes experiencing deeper ducts) and according to seasonal, geographical, and diurnal conditions. Land-based ducts of this type are normally short lived. Advection is another process that leads to the formation of ducts. The advance of warm dry continental air over a cool surface, such as wet ground or the cool sea, causes a cooling and moistening of the lower layers and results in a temperature inversion and steep negative humidity gradient. The warmer and drier the air, the stronger the gradient. Advection ducts can also sometimes be observed when warm, moist air is advected over a cooler sea, resulting in a temperature inversion. So-called quasi-advection, where a strong cool wind blows over a warm wet surface, such as the sea, can also lead to a steep negative humidity gradient proportional to the strength of the wind. These advection ducts are particularly important in coastal regions. Frontal weather processes, such as an advance of cool surface air lifting a stable warm air mass can lead to a temperature inversion, with an elevated duct resulting. These ducts also are usually short lived. Anticyclonic subsidence is the most common reason for elevated ducts in midlatitudes where a temperature inversion is caused by the subsiding air. In particular, when a stratocumulus cloud forms the boundary layer beneath the anticyclonic inversion, ducting can occur due to the hydrolapse. In some regions where stratocumulus clouds are semipermanent, such as in the subtropics, semipermanent ducting can be prevalent. These ducts can form up to 3 km above the Earth's surface. It is only the ducts in the first few hundred meters of the ground that are of any interest to point-to-point microwave radio link designers. Nocturnal radiation of heat with subsequent cooling of the Earth's surface under clear skies, can lead to the formation of a temperature inversion. Dew deposition on the ground leads to an increase of humidity with height, which reduces the possibility of a duct formation. On the other hand, if the ground is wet due to dew deposition, a duct could form at sunrise as the dew evaporates. Light surface winds tend to reduce the dew deposits (and hence reinforce the hydrolapse) and can result in severe cooling of the Earth's surface (thus reinforcing the temperature inversion). If the wind is too strong, a mixing of the atmospheric layers occurs and a duct will not form [5]. Temperature inversions are the primary cause of ducts, which influence radio link propagation. As pollution gets trapped in the duct, these ducts are often visible. The various synoptic processes that lead to abnormal refraction conditions are illustrated in Figure 5.5.

One may read in various texts written in North America and Western Europe that multipath fading is a summer phenomenon—this is probably related to the high humidity and dew content in summer. In temperate climates where the surface heating is shallow, the problem appears to be worse in winter. A small amount of moisture from dew deposits can lead to a steep humidity gradient even though the actual humidity level is low. This coupled with a temperature inversion can result in severe multipath fading conditions.

5.1.7 Modified Refractivity

If one considers super-refractivity, when the gradient becomes more negative, the signal propagation range is increased due to refraction. At a certain critical

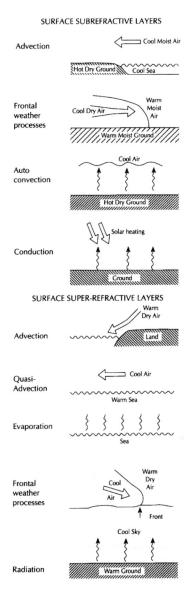

Figure 5.5 Synoptic processes causing abnormal conditions.

gradient ($G = -157$), the ray would be parallel to the Earth and have a theoretically infinite range. After this, if the gradient becomes more negative, blackout conditions can occur because the radius of the radio beam becomes less than the radius of the Earth and be so strongly refracted downward that it does not reach the receive antenna. This condition, where the refractivity gradient is less than −157, has already been described as a ducting condition.

For various applications such as ray tracing, a modified refractivity called the refractive modulus is used when considering ducting. This is defined as

$$M = N + 10^6 \, h/a \qquad (5.11)$$

where h is the height above the Earth's surface and a is the radius of the Earth. This transformation allows one to consider the Earth as flat with the atmospheric condition having the characteristic of M. Substituting the Earth's radius (6371 km) in (5.11) yields

$$M = N + 157 \cdot h \qquad (5.12)$$

where h is in kilometers. Hence,

$$dM/dh = G + 157 \qquad (5.13)$$

The gradient of the refractive modulus with height (dM/dh) is zero when the refractivity gradient $G = -157$ and is negative for ducting conditions, so it can be seen why it is useful for analyzing ducting conditions. In practice there are three forms of the M-profile under ducting conditions. The first is where the duct has a negative M gradient from the surface to the top of the duct. The second is also a surface duct but a positive gradient of M is present at the surface, which means that the trapping layer does not extend right down to the Earth's surface. This happens because the value of M at the surface is greater than the value at the top of the duct, forming an S-shaped surface duct. The third type is where the value of M at the surface is less than the value at the top and so the duct does not extend down to the ground but is elevated. Its depth extends from the local minimum to the height at which the M-value equals that at the top of the duct. This is called an elevated duct. The M-profile ducting diagrams are shown in Figure 5.6.

Under ducting conditions the individual ray emitter from a microwave antenna can cross at some point and cause interference. This results in an uneven distribution of power in space, where in certain areas there is a concentration of energy but at others the signal level is low. For microwave frequencies it is convenient to use geometric optics to analyze the problem. Computer-based ray-tracing techniques are therefore used to analyze this phenomenon. A typical ray-tracing curve under ducting conditions is shown in Figure 5.7.

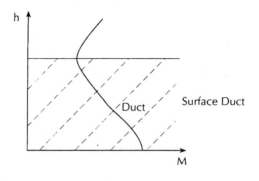

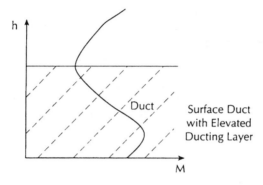

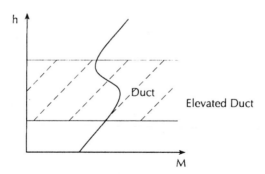

Figure 5.6 *M*-profile ducting types.

5.2 Free-Space Propagation

Radio waves are affected by the presence of the Earth and the atmosphere surrounding it. For microwave point-to-point links it is the nonionized lowest portion of the atmosphere (below roughly 20 km), called the troposphere, that

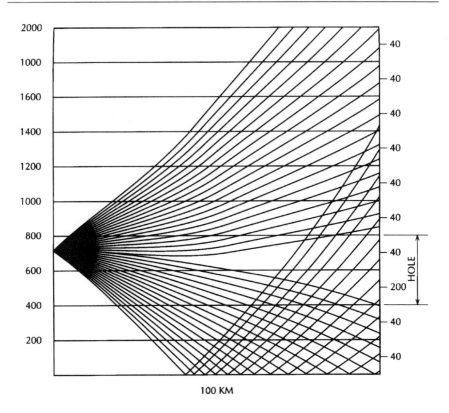

100 KM

Figure 5.7 Ray tracing in ducting condition.

is of interest. For path-planning purposes it is useful to define a reference position where the propagation can be considered unaffected by the Earth. The loss between two antennas unaffected by the Earth is called the free-space loss and can be derived as follows.

Consider a point source with power Pt. If one imagines that this is bounded by a sphere, with a radius d, the power ratio between transmitting source and receiving antenna at point d can be expressed as

$$Pt/Pr = (4\pi d/\lambda)^2 \tag{5.14}$$

Now if we substitute $\lambda = c/f$ into (5.14) we get

$$Pt/Pr = (4\pi df/c)^2 \tag{5.15}$$
$$= (4\pi/c)^2 + d^2 + f^2$$

Converting to decibels we get

$$10 \log(Pt/Pr) = 10 \log((4\pi/c)^2 + d^2 + f^2) \text{ dB} \qquad (5.16)$$

If one now assumes that two isotropic radiators are used as source and sink antennas, then the free-space loss between the two antennas can be expressed as

$$
\begin{aligned}
FSL &= 10 \log(Pt/Pr) \text{ dB} & (5.17) \\
&= 20 \log(4\pi/c) + 20 \log d \text{ (m)} + 20 \log f \text{ (Hz)} \\
&= 20 \log(4\pi/c) + 20 \log d \text{ (km)} - 20 \log 10^3 \\
&\quad + 20 \log f \text{ (GHz)} - 20 \log 10^9 \\
&= 20 \log(4\pi) - 20 \log(3 \times 10^8) - 20 \log 10^3 - 20 \log 10^9 \\
&\quad + 20 \log d \text{ (km)} + 20 \log f \text{ (GHz)} \\
&= 92.4 + 20 \log d_{km} + 20 \log f_{GHz} \text{ dB} & (5.18)
\end{aligned}
$$

Note that a doubling of the distance represents a 6-dB decrease in the signal. This is an important consideration when frequency interference is considered. If one were to transmit a microwave signal over a path of fixed length, under normal conditions the only path attenuation is the free-space loss as expressed by (5.18) and the receive level recorded at the distant end is referred to as the nominal receiver level.

5.3 Power Budget

In order to determine the performance of a link, one has to work out the percentage of time the signal will be below the threshold of the radio receiver relative to the total time period. One therefore has to determine the difference between the nominal signal level and the receiver threshold level. This is known as the fade margin. It is imperative to be able to predict the expected receive level of a radio link for two main reasons. Specifically, one must ensure that an adequate fade margin exists in the design phase and one needs to know when the antennas have been panned correctly during the commissioning phase. Adding the various gains and losses over the path from transmitter output module to receiver demodulator input is called the power budget.

5.3.1 Receiver Threshold

The receiver threshold is the minimum signal required for the demodulator to work at a specific error rate. Two thresholds are normally defined, one at a BER of 10^{-6} the other at a BER of 10^{-3}.

The receiver threshold is dependent on the minimum S/N required at the receiver input, the noise figure of the receiver's front-end, and the background thermal noise (Pn)

$$Pn = kTB \qquad (5.19)$$

where k is the Boltzman's constant (1.38×10^{-23}), T is the temperature in Kelvins, and B is the bandwidth of the receiver. Microwave radio manufacturers will specify the receiver threshold values of the radio equipment relative to the bandwidth of the system.

5.3.2 Nominal Receive Level

Under unfaded conditions the link budget is

$$P_{RX} = P_{TX} - L_{TX} - FL_{TX} + A_{TX} - FSL + A_{RX} - FL_{RX} - L_{RX} \qquad (5.20)$$

where P_{RX} is the unfaded receive level in dBm (nominal RX level), P_{TX} is the transmitter output power in dBm, $FL_{TX,RX}$ is the feeder loss of cable or waveguide in decibels, $A_{TX,RX}$ is the antenna gains in dBi, FSL is the free-space loss in decibels, and $L_{TX,RX}$ is the branching losses.

5.3.3 Fade Margin

The difference between the nominal receive level and the receiver threshold level is available as a safety margin against fading. For this reason it is known as the fade margin. It will be shown later that each hop can be designed with different fade margins in digital systems, unlike analog systems that were designed to a specific fade margin (usually 40 dB). The fade margin to be achieved should match the availability and performance objectives set.

5.4 Fading on Microwave Links

A microwave radio link suffers various signal fluctuations in time for a number of reasons. These signal variations around the nominal receive level are commonly referred to as fading. If one were to plot the receive AGC level with time, one would find that there is a constant variation in the receive level. This is due to the small variations in the point index gradients of refractivity through which the signal will pass over the path. This type of fading is called scintillation and has no effect on the overall system design and can be ignored. Seasonal or diurnal variations in the average refractivity gradients do affect the signal and cause refractive fading. This includes diffraction fading; beam spreading (or defocusing); multipath fading, which results in Raleigh fading in narrowband systems and Raleigh plus selective fading in wideband systems; and ducting or blackout fading. Another major cause of fading is rain attenuation in high-frequency systems (i.e., mainly above 10 GHz). At this stage it may be helpful to clarify a term that is often misused in the industry, namely, flat fading. Strictly speaking, flat fading is nonfrequency dependent and therefore results in an equal attenuation across the bandwidth of the receiver. Rain fading and diffraction fading are examples of true flat fading. This term is also used, however, to describe the attenuation from multipath fading in narrowband systems, which exhibits a Raleigh probability distribution curve. Multipath fading is caused by two or more signals that have traveled slightly different paths and therefore add or cancel at a particular frequency depending on their phase relationship. Clearly, they would have a slightly different phase relationship at another frequency within the bandwidth of the receiver. In a narrow bandwidth system (e.g., 2 Mbit/s), the receiver will generate errors as the signal level reduces due only to increased thermal noise. It is as though there was an even amplitude reduction across the bandwidth of the receiver; hence, it is termed flat fading. In wideband systems, this same multipath effect causes not only amplitude reductions but also in-band distortion, which results in errors that are not amplitude dependent. This is termed selective fading or dispersive fading.

Having set the scene, let us consider some of these fading mechanisms in a little more detail.

5.4.1 Atmospheric Absorption Including Rain

The main elements in the atmosphere that absorb electromagnetic energy are water vapor and oxygen. Oxygen resonance occurs at about 0.5 cm (60 GHz) and water vapor resonance occurs at 1.3 cm (23 GHz). For frequencies below 5 GHz, the effect is thus negligible; up to 10 GHz, the rain attenuation is normally insignificant compared to the attenuation caused by refraction effects. Water vapor absorption and rainfall attenuation are thus usually only considered above 10 GHz. Rain also causes the signal to scatter, but this effect is negligible for a point-to-point radio link. As the rain rate increases the instantaneous amount of water in the path increases, resulting in very high attenuation. The higher the frequency, the higher the water absorption. Higher frequency bands, such as 23 GHz and 38 GHz, are thus only useful for short hops. The absorption has both positive and negative affects. On the negative side, rain attenuation severely limits link lengths; however, at 23 GHz for example, the extra attenuation is used to improve frequency re-use. Absorption effects due to mist, fog, snow, and dust are negligible compared to rain attenuation. The water vapor density curves are provided by the ITU [6]. Rainfall rates are provided by the ITU in terms of different zones defined as the rainfall rate, which is only exceeded for 0.01% of the time. In order to obtain this information it is necessary to use high-speed tipping buckets to accurately measure the rainfall rate. In order to compare rainfall rates obtained from a weather bureau, the same integration period must thus be used for a meaningful comparison.

The path attenuation is the sum of the attenuation due to atmospheric gases (including water vapor) and the attenuation from rainfall

$$A_{dB} = \gamma_a d + \gamma_R d \qquad (5.21)$$

where γ_a is the specific attenuation in dB/km obtained from the ITU [7], γ_R is the specific attenuation in dB/km obtained from the ITU [8], and d is the path length (km).

5.4.2 Diffraction Fading

Under conditions where a positive refractivity gradient is present, the radio beam is refracted upward and hence the portion of the wavefront that is received at the distant end has traveled closer to the ground than usual. With certain terrain profiles and depending on the height of the two antennas, this could result in loss of visibility and a resulting diffraction loss. The radio path designer needs to ensure that this diffraction loss does not cause an outage in excess of the availability objectives by placing the antenna at suitable heights above the ground. In order to determine what that suitable height is we need to first understand the diffraction mechanism.

5.4.2.1 Fresnel Zones

In order to ensure that the diffraction loss due to loss of visibility is acceptably small under all possible atmospheric k variations, it is essential that a path profile be drawn that shows the antenna heights and terrain heights adjusted to account for the k variation. Normally the particular area of the profile that has the least clearance from a direct ray drawn between the two antennas is called the dominant obstacle. Note that in some cases the dominant obstacle can be different for different values of k. When the LOS ray between the two antennas just clears the dominant obstacle, the condition is referred to as grazing LOS. Because the microwave beam travels as a wavefront and not as a single ray, grazing LOS will not result in free-space propagation conditions.

In other words, even if the other end is clearly visible, this is not sufficient to ensure that obstruction losses will not occur. To determine the amount of clearance necessary for this condition, the concept (developed in geometric optics) of Fresnel zones (pronounced Fray-nel) is useful. This concept is based on introducing an infinitely thin screen into the direct path of an electromagnetic wavefront and analyzing the bending around the screen, which is called diffraction. According to Huygen's principle (pronounced Hoy-gen), a wavefront can be considered as an infinite number of secondary wavelets each radiating energy in the same way as the primary source, thus producing a secondary wavefront. This is illustrated in Figure 5.8.

The received field strength that would be measured at the receiving antenna is thus the sum of an infinite number of tiny wavelets produced by the transmitting antenna. Elements off the main axis of the microwave beam, forming the direct path between the two antenna apertures, thus contribute to the overall receive field; if they are blocked, the receive field strength at the antenna is affected. Obviously in a practical sense, at a certain distance from the axis these components have a negligible effect. The phase delay of the off-axis components is also critical. For example, if the off-axis component is delayed by half a wavelength, then when they are summed at the receive side they will cancel; if the difference was a full wavelength, they would add. Under free-space conditions this complex phase arrangement results in the normal

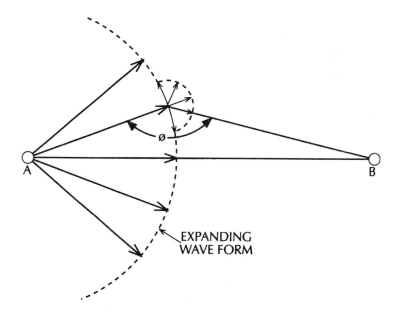

Figure 5.8 Huygen's principle.

condition of the field strength decreasing with the square of the distance from the antenna. However, if a screen is introduced that blocks off a portion of the wavefront, the pattern looks quite different and areas of minima and maxima can be observed as the screen blocks more and more of the wavefront. This is seen visually in optics where a screen with a hole in it blocks a light source, resulting in a diffraction pattern of light and dark concentric circles. The boundary points are where the path difference between the direct ray and a ray from a secondary wavelet are multiples of half-wavelengths. According to the theory of Fresnel zones, these points are defined by a series of ellipses where the two antennas are at the loci. This is illustrated in Figure 5.9.

Provided one is in the far-field ($2D^2/\lambda$, where D is the antenna aperture in the same units as wavelength), the first Fresnel radius F1 can be expressed as

$$F_1 = \sqrt{(\lambda (d_1 \cdot d_2)/(d_1 + d_2))} \qquad (5.22)$$

where λ is the wavelength of signal in meters, d_1 is the distance of the obstacle from transmitter A in meters, and d_2 is the distance of the obstacle from transmitter B in meters.

Using more convenient units we can express (5.22) as

$$F_1 = 17.3\sqrt{(1/f(d_1 \cdot d_2)/(d_1 + d_2))} \text{ meters} \qquad (5.23)$$

where f is the frequency in gigahertz and d_1, d_2 are in kilometers.

The subsequent nth Fresnel zone can be determined as

$$F_n = F_1\sqrt{(n)} \qquad (5.24)$$

where n is an integer.

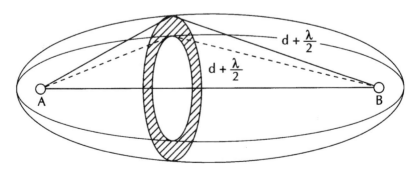

Figure 5.9 Fresnel zones.

At the midpoint the radius will be at its maximum. Equation (5.24) can thus be adjusted as

$$F_1 \text{ (midpoint)} = 17.3 \sqrt{(d_1/2f)} \qquad (5.25)$$

It is interesting to note that the difference in path length between one Fresnel zone and the next is only one-half a wavelength by definition. At gigahertz frequencies this is only millimeters; yet on an average path of 50 km, this represents tens of meters of clearance difference.

One major deduction to be made from the theory of Fresnel zones is that, in theory, provided at least 60% of the first zone is clear of any obstruction, the effect of the Earth can be ignored and the path loss approximated by the free-space loss. This is shown in Figure 5.10.

One needs to treat this with caution, however, because it is important to realize that this is a theoretical concept and has certain practical limitations. For example in practice, the two sources are not omnidirectional point sources but high-gain directional microwave antennas with a certain finite diameter usually of a few meters or more. Further, the infinitely thin screen that should block the wavefront in practice is usually a mountain or similar obstacle that is very rough compared to the wavelength of propagation and only forms a screen below the main axis. Despite this, there are some practical deductions that can be made about diffraction loss, and this subject is adequately treated by the ITU [9]. The various formulas required to calculate diffraction loss for knife edge and isolated obstacle, for example, are covered. The ITU [10] provides the formula for diffraction loss over average terrain as

$$A_d = -20h/F1 + 10 \text{ dB} \qquad (5.26)$$

where h is the height above the Earth's surface and $F1$ is the first Fresnel Zone radius.

One needs to be careful when considering the theoretical curves for diffraction loss as to whether they are produced for a screen with a hole in it (concentric circles) or whether it is a horizontal screen because although the shape of the response is the same, the values are different. Different methods are appropriate for different types of physical obstructions, such as knife edge, isolated rounded obstacle, smooth earth, and irregular terrain [11]. The planner must choose the most appropriate method for the particular obstruction. In practice, a few relevant methods should be used and the worst case values chosen to reduce the risk of the theoretical models producing optimistic results. In is unlikely, for example, that a knife edge actually exists in practice. A grazing LOS diffraction loss of 6 dB is thus highly optimistic for most paths.

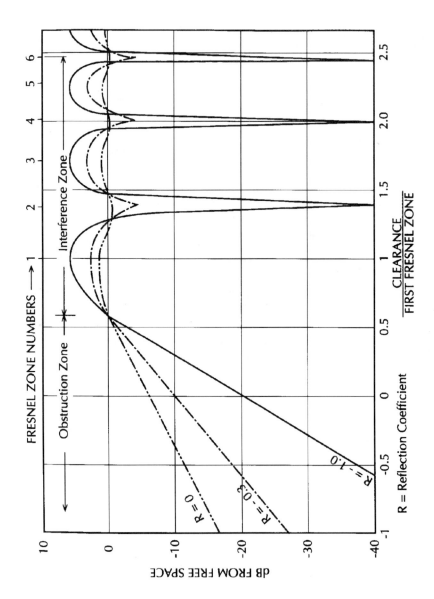

Figure 5.10 Attenuation versus path clearance.

5.4.2.2   Minimum *K* Values

As already stated, the condition where diffraction losses are considered is when steep positive gradients of refractivity are present. This corresponds to the minimum *k* value. This minimum value varies from region to region and also from season to season. Obviously the higher the minimum value of *k*, the lower the required tower height. In order to minimize costs, attempts have been made to lower the tower height and accept the resulting diffraction loss, just ensuring that the loss does not result in an outage. However, the minimum *k* condition is due to relatively stable atmospheric conditions (which could prevail over periods of hours rather than seconds or minutes). If one does not have a fully clear path but has accepted a certain diffraction loss for low *k* values, it is very dangerous to underestimate the minimum *k* value. Stated differently, unless the minimum *k* condition can be accurately determined, preferably from field results, one should ensure that only negligibly small diffraction losses are possible. It is possible to calculate *k*min values from field data using a method developed by Mojoli [12] that is adequately described in the reference. It should be noted, however, that this method assumes the existence of reliable and accurate point refractive index gradient data, and the absence of this data could lead to erroneous *k*min estimates. It is also important to remember that the *k*min estimate is based on averaging the point refractive index gradients over the path length, so the particular gradient at the dominant obstacle may be less than the kmin estimate even if the estimate was correct. Further, for small proportions of the time, localized areas along the path that may have extreme gradients could result in beam divergence, thus invalidating the previous method. (The method is based on summing the point gradients to achieve a composite gradient.)

5.4.2.3   Clearance Rules

It is important to position the antennas at an appropriate height that will protect the signal from diffraction loss under all possible *k*-factors. It has been shown that if the antennas are too high they risk second Fresnel zone clearance, which could result in very high signal attenuation. This also increases the amount of interference experienced and can lead to a demand for taller towers, which dramatically increases costs. In order to ensure negligibly small diffraction losses under both *k* extremes, it is traditional to use clearance rules. A commonly used rule on analog systems is to have full Fresnel zone clearance at a median *k* value of 4/3 and 60% Fresnel zone clearance under a minimum *k* value of 2/3.

Special *k* paper with curves at these values were used to draw the path profiles and work out the required clearances. PC-based software programs have replaced this manual procedure.

Clearance rules must be used with care because they have three basic shortfalls. First, they do not take into account the frequency band used. The clearance risk increases as the frequency increases due to the reduction in the size of the first Fresnel zone. Second, they do not take into account the required performance criteria. They normally assume that no diffraction loss can be tolerated. Third, the two criteria often conflict with one another. The trend has been to place the antennae at the highest height dictated by the two criteria, thus increasing the risk of second Fresnel zone clearance and resulting in unnecessarily high towers. Earlier applications of clearance rules based on the 2/3- and 4/3-ruled paper resulted in antenna heights being unnecessarily high. The 2/3 (0.67) value for minimum k was based on a 30-km-long path measured in France. This value has tended to be used throughout the world on various paths regardless of path length. It is well known that minimum k values increase with path length due to the averaging effect of the point gradients of refractivity across the path. The ITU has produced a curve that shows a conservative estimate of minimum k values with hop length [13] (see Figure 5.11).

This curve shows that a 30-km path would have a minimum k value of 2/3 whereas for a 50-km path it would be 0.8. Using this minimum k value with the traditional clearance rules makes a significant difference to antenna

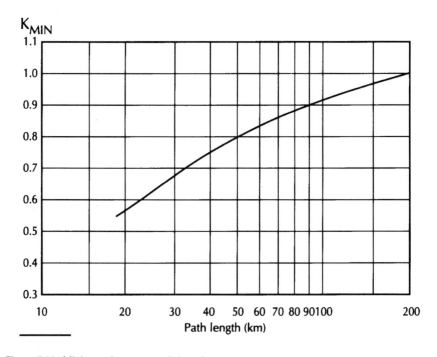

Figure 5.11 Minimum k versus path length.

height requirements. Where the median value of k is available, this should be used instead of the 4/3 value. This value can be reliably calculated from meteorological data. Most countries will have a database of refractivity values taken using radar soundings and weather balloons for weather predictions. These soundings are usually taken at least twice a day at key locations and measure the pressure, humidity, and temperature at different heights above the ground. The refractivity gradient can easily be calculated from this data. Unless these measurements are taken often enough to include extreme values of gradients, it is the author's opinion that they cannot be used to predict minimum k values. Median k values, on the other hand, will have sufficient data to be statistically reliable.

Fresnel zones are normalized to be independent of frequency. It is important to remember that the clearance is not independent of frequency. The radius of the Fresnel zones are inversely proportional to the square root of frequency; therefore, as the frequency increases the radius becomes smaller and clearance is easier to attain. This reduced clearance also increases the risk of obstruction fading. The Earth bulge is independent of frequency; thus, for a given Earth bulge the intrusion into the first Fresnel zone is greater for higher frequency links. Lower frequency links tend to be more reliable. This is why the 2-GHz band was chosen for the world's longest radio link, which spans more than 300 km across the Red Sea. Very high frequency links, such as 23 GHz, have such a small Fresnel zone that it makes more sense to ensure clearance with a safety margin of say 5 meters than to work out a percentage clearance of the first Fresnel zone.

Network planners have tended to be extremely conservative when it comes to clearance rules due to the fear of a diffraction fade outage. The rules tend to ensure that absolutely no diffraction loss would be experienced. Diffraction loss is a slow fading event, which means that if the fade margin were exceeded, the link could be out for some minutes if not hours. It should be remembered however, that diffraction fading is caused by low k values and multipath fading by high k values. The probability of multipath fading being superimposed on diffraction fading is thus very low. To allow a small amount of diffraction fading is thus a low risk. Providing the path profile information, including ground clutter, is accurate. It makes more sense to position the antennas at the required height specified by the first clearance rule and then to ensure that under minimum k conditions the diffraction loss does not exceed the fade margin. The diffraction loss can be calculated using the methods discussed earlier. To build in a safety factor for any unexpected conditions, instead of utilizing the entire fade margin for diffraction loss, one could choose 25% of the fade margin as the limit. This will have a significant effect on the tower height requirement yet still limit the risk of outage to 25% of the overall fading allowed. The previous method ensures that the antennas are optimally placed for the median condition, which is what the system will experience for the maximum time, and significantly reduces the tower height requirements. On a large project the cost savings can be enormous.

The preceding discussion has ignored clearance criteria for space diversity systems. Space diversity is usually applied to overcome the effects of multipath fading associated with high k values and thus the antennas are not required during obstruction fading. The traditional approach is to set the antenna position of the main antennas and diversity antennas using the usual clearance rules. In some cases this may have been relaxed to only ensure adequate clearance of the space diversity antennas under median k conditions only. This would still have often impacted on the height requirement of the main antennas to achieve adequate spacing between main and diversity antennas. Considering that diffracted links are more immune to multipath fading it makes sense to have the diversity antennas diffracted while maintaining the main antenna heights as originally calculated. This combination ensures that the diversity antennas give good multipath protection and, during low k diffraction conditions, the main antennas provide protection against obstruction fading. Apart from the obvious performance improvement, significant cost savings can be made using lower towers and shorter feeder runs. Interference is also reduced.

5.4.3 Refractive Fading

In the introduction the point was made that changes in the refractive index profile not only caused diffraction loss problems but also resulted in beam spreading, multipath propagation, and ducting. Diffraction loss is associated with steep positive gradients of refractivity, the preceding effects are caused by steep negative gradients of refractivity. Strictly speaking, an atmospheric layer that exceeds (more negative than) -157 N-units/km, is called a duct since this is the boundary condition at which the signal is exactly parallel to the Earth. Gradients that are more steeply negative than this will trap the microwave signal within this duct region. Experience has shown that negative layers that exceed -100 N/km will cause multipath, hence the term duct has been extended to include this region. A duct in the context of a multipath fading discussion is an atmospheric layer that is more negative than 100, but a true ducting condition is where the gradient exceeds -157 N/km.

5.4.3.1 Beam Spreading

When a super-refractive atmospheric condition occurs, a caustic can be created along which focusing of the radio beam energy occurs, with so-called radio holes created around it. This is illustrated in Figure 5.12.

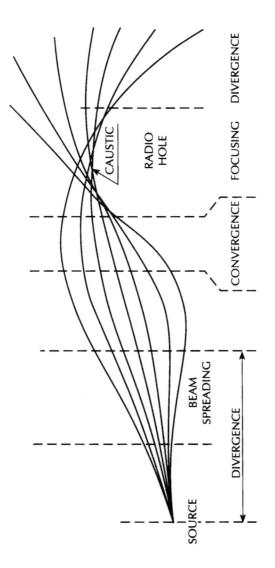

Figure 5.12 Effect of ducting layer on microwave rays.

Depending on the height of this layer and the height of the transmit and receive antennas, a signal enhancement or reduction can occur. As this condition can be relatively stable (last hours rather than minutes or seconds) the fading is slow and results in a mean depression (or enhancement) of the signal. This fading is less severe than multipath fading, and it is not frequency selective because it is a single path phenomenon. Ray-tracing techniques have shown that it is most pronounced when the transmit and receive antennas are at the same height with the center of the abnormal layer just below the path. Provided the mean depression is not severe, the main affect this problem has on the radio system is that the mean level from which multipath fading occurs is reduced, thus increasing the multipath outage. Fading formulas are empirically based; hence, it is the author's view that for most links this affect is already included in the outage predictions.

5.4.3.2 Multipath Propagation

An atmospheric duct has a very steep negative gradient and tends to strongly refract the radio signal downward. Depending on where the duct is situated relative to the radio signal, more than one transmission path is possible. This condition is termed atmospheric multipath. (Multipath can also occur from ground reflections, and in theory a subrefractive layer below the two antennas could also result in multipath conditions.) This type of fading can result in very deep fades that are frequency selective. In wideband radio systems the frequency-selective nature of multipath fading is very important, so much so that it will be treated separately. In practice, because the slow fading associated with beam spreading and the fast multipath fading are associated with the same atmospheric condition, the two occur simultaneously with very deep fast fades superimposed on a slow mean depression of the signal. Multipath fading is the main fading effect of radio link systems operating below 10 GHz. The secondary signal that causes the multipath can result from either an elevated duct that bends a part of the signal, which is normally lost in space, back to interfere with the main signal or a ground reflected signal as shown in Figure 5.13.

In the past it was thought that most multipath outages were the result of interference from elevated ducts; however, it is widely understood nowadays that it is actually nonspecular ground reflections interfering with a defocused main signal that is the dominant multipath fading effect. The empirical fading models used to predict fading outages reflect this fact in that the terrain profile has a dominant effect on the overall outage. In the previous version of ITU 530 (version 6) the grazing angle—that is, the angle of the ground reflection—was included; in the latest version, however, it was dropped due to the difficulty of accurately calculating it from rough paths.

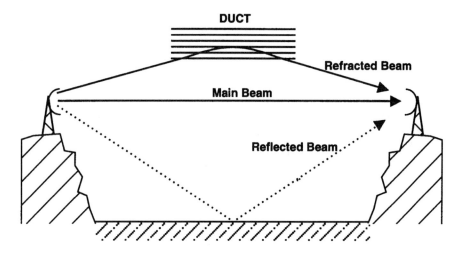

Figure 5.13 Elevated duct.

5.4.3.3 Ducting

In the presence of a ground-based duct the refractivity gradient is so strongly negative that the beam is bent downward in such a way that it is not received at the far end. This condition is often termed blackout or ducting. In areas where the mean variation of refractivity gradients is very high, it presents a real risk to successful propagation because the fade usually exceeds any reasonable margin one could include in the design phase. The heights of the antennas are critical and the usual method to overcome this problem is to analyze the risk involved for each antenna height and to place the antennas at such a height so as to minimize the risk of blackout. The ITU provides refractive index data that helps to quantify the risk of this type of fading occurring. Any areas where the negative gradients exceed 50% occurrence are obvious high risk areas—this includes Saudi Arabia where semipermanent ducting conditions occur.

References

[1] Bean, B. R., and E. J. Dutton, *Radio Meteorology*, New York: Dover Publications Inc., 1966.

[2] Boithias, L., and J. Battesti, "Protection Against Fading on Line-of-Sight Radio-Relay Systems," *Ann. des Telecomm.*, Sept.–Oct. 1967. (French)

[3] ITU-R PN.310-9, Geneva, 1994.

[4] ITU-R P.453-6, Geneva, 1997.

[5] Allen, E. W., *Correlation of Clear-Air Fading with Meteorological Measurements*, Rockwell, 1988.

[6]   ITU-R P.836-1, Geneva, 1997.

[7]   ITU-R P.676-3, Geneva, 1997.

[8]   ITU-R P.838, Geneva, 1992.

[9]   ITU-R P.526-5, Geneva, 1997.

[10] ITU-R P.530-7, Geneva, 1997.

[11] Rice, P. L., K. A. Norton, and A. P. Barsis, et al., "Transmission Loss Predictions for Tropospheric Communication Circuits," NBS Publ., Technical Note 101, Jan. 1967.

[12] CCIR Report 718-3, Geneva, 1990.

[13] CCIR Report 338-6, Figure 2, Geneva, 1990.

6

Antenna Considerations

The main component under the control of the radio planner in terms of detailed link design is the antenna. Its general characteristics—including gain, interference rejection, height above the ground, and tower loading—are all critical factors in a successful design. For this reason a full chapter has been devoted to the antenna and its associated feeder and ancillary system.

6.1 Electromagnetic Theory Fundamentals

When analyzing electrical circuit parameters one usually analyses the effect of a current flow through a certain impedance that sets up a voltage across the element. When considering antennas it is more useful to analyze the effects in terms of the electric and magnetic field vectors since the wavefront, which is generated by an antenna, travels as an electromagnetic field. An electromagnetic (EM) wave has two orthogonal components: an electric field and, perpendicular to this, a magnetic field. These two fields can both be considered sinusoidally varying signals perpendicular to each other, both normal to the direction of propagation, as shown in Figure 6.1.

In an EM wave, the electric and magnetic fields interact with one another. A changing magnetic field will induce an electric field and a changing electric field will induce a magnetic field. The wavefront is the imaginary line that is drawn through the plane of constant phase. A plane wave is one that has a plane wavefront. A uniform wavefront is one that has constant magnitude and phase. The magnitude of the electric and magnetic field vectors are thus equal and occur "in phase." The peaks and troughs thus occur at the same time for both waves. The EM wave travels in a direction perpendicular to the wavefront.

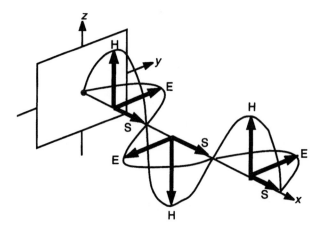

Figure 6.1 Electromagnetic wavefront.

EM waves in free space travel as a uniform plane wave; this is known as a transverse EM (TEM) wave.

6.1.1 Period

The period of the wave is the length of time before the wave repeats itself and can be expressed as

$$T = 1/f \qquad (6.1)$$

where f is the frequency (the number of cycles in 1 sec, measured in Hertz).

6.1.2 Wavelength

The wavelength of the signal is the distance between two points of the same phase, which depends on the medium in which it is traveling, and can be expressed as

$$\lambda = v/f \qquad (6.2)$$

where v is the velocity of propagation (m/s) and f is the frequency (Hertz).

6.1.3 Velocity of Propagation

The velocity of propagation can be expressed as

$$v = 1/\sqrt{(\epsilon \cdot \mu)} \qquad (6.3)$$

where ϵ is called the permittivity and can be expressed as $K \cdot \epsilon_0$; ϵ_0 is the permittivity of free space = 8.854×10^{-12} F/m; K is the relative permittivity, for example, $K(\text{air}) = 1$; μ is the permeability and can be expressed as $K_m \cdot \mu_0$; μ_0 is the permeability of free space = 1.257×10^{-6} H/m; and K_m is the relative permeability dependent on the material used, for example, K_m (aluminum) = 1.00000065.

All dielectric mediums are specified in terms of permittivity and permeability, which are measures of capacitivity and inductivity, respectively.

In a vacuum,

$$v = 1/\sqrt{(\epsilon_0 \cdot \mu_0)}$$
$$= 3 \times 10^{-8} \text{ m/s} \tag{6.4}$$
$$= c \text{ (speed of light)}$$

The velocity of propagation in other mediums depends on the permittivity and permeability of that medium. In air, the permittivity and permeability is approximately the same as in a vacuum, so the microwave beam travels at the speed of light, irrespective of frequency.

6.1.4 Polarization

The polarization of the signal corresponds to the plane of the electric field vector. If one imagines a sinusoidal wave traveling perpendicularly out of the page, the amplitude vector would swing from a positive maximum through zero to a negative maximum. In this plane, the electric vector oscillates vertically and thus is vertically polarized.

6.1.5 Power Density

An EM wave transports energy that can be represented as a power density P_d in Watts/m^2. A point source of radiation that transmits energy uniformly in all directions is called an isotrope. If one considers a sphere around this isotropic source with power output P_t as having an area $A = 4\pi r^2$, then the power density (P) can be expressed as

$$P = P_t/(4\pi r^2) \tag{6.5}$$

6.2 Antenna Characteristics

An antenna is basically a radiating element that converts electrical energy in the form of current into an EM wavefront in the form of oscillating electric and magnetic fields. Any current flow in a conductor will set up a magnetic field. Any varying current flow will generate a varying magnetic field that in turn establishes an electric field. There is thus an interaction between electric and magnetic fields that results in EM propagation. The faster the variation of these fields, the greater the radiation from the conductor element (antenna). The field components that are not radiated make up the capacitive and inductive parts for the antenna, resulting in the antenna exhibiting a complex impedance rather than just resistance.

6.2.1 Gain

An antenna is a passive device and, thus by definition, cannot amplify the signal; however, it can shape the signal to be stronger in one direction than another. Consider a balloon: If one were to squash the balloon at the sides, it would expand at the ends. This is the basis of what we consider as antenna gain. The reference is an isotropic radiator that by definition has zero gain. The amount by which the antenna shapes the signal in a particular direction is described in terms of its gain. When one talks of the gain of the particular antenna, one is referring to the gain at the boresight of the antenna, that is, the radiation coming directly out of the front of the antenna. The gain is expressed as the ratio between the reference power density (P) of an isotropic radiator and the power density in the particular direction that one is considering. This is usually expressed in a logarithmic scale in decibels. Microwave antennas are usually specified in dBi and can be expressed as

$$\text{dBi} = 10 \ \log_{10} \ P/P_{\text{di}} \tag{6.6}$$

where P is the power density in the direction considered and P_{di} is the power density of an isotropic radiator. Since an isotropic radiator cannot be constructed in practice, another way of expressing gain is to compare it to a folded dipole, which is the closest physical antenna construction to the mythical isotropic radiator. This is typically done for VHF and UHF antennas, which are expressed as dBd

$$\text{dBd} = 10 \ \log_{10} \ P/P_{\text{dd}} \tag{6.7}$$

where P is the power density in the direction considered and P_{dd} is the power density of an omnidirectional dipole. The gain of an antenna expressed in dBd is 2.16 dB less than relative to an isotropic antenna (dBi).

For microwave antennas, the gain is dependent on the area of the antenna aperture. The gain of an antenna can be expressed as

$$G \text{ (dBi)} = 10 \log \eta \left(4\pi A_a / \lambda^2\right) \tag{6.8}$$

where η is the efficiency of antenna aperture, A_a is the area of antenna aperture, and λ is the wavelength of the signal.

Maximizing the gain by illuminating the entire parabolic face of the reflector would result in a very poor front-to-back (F/B) ratio. This is defined in Section 6.2.3. The illumination area is thus purposely reduced to improve the sidelobes and back lobes. These are defined in Section 6.2.2. Typical efficiencies range between 50% and 60%. If one assumes a parabolic dish with an efficiency of 55% and expresses the units in meters and gigahertz, a useful expression results

$$G \text{ (dBi)} = 17.8 + 20 \log \left(d \cdot f\right) \tag{6.9}$$

where d is the antenna diameter (meters) and f is the frequency of horn feed (gigzahertz). This is a very useful formula for the radio planner to use to estimate the gain of any parabolic microwave antenna if the exact details of the actual antenna are not available.

6.2.2 Sidelobes

Microwave antennas are intended to be directional. The maximum radiation is thus in the direction of propagation. In practice, it is impossible to shape all the energy in this direction. Some of it spills out off the sides and back of the antenna. Due to the complex phases set up in an antenna pattern, lobes result. The main lobe is around the center of the antenna. Sidelobes of lesser amplitude result around the rest of the antenna. The aim of a directional antenna is to maximize the energy in the main lobe by minimizing the energy in the sidelobes. It is important to understand the radiation patterns when panning antennas to make sure that one does not pan the signal onto a sidelobe.

6.2.3 Front-to-Back Ratio

As already discussed, not all the energy radiates out the front of the antenna. Some of it radiates out of the back lobe. The F/B ratio is defined as the ratio of the gain in the desired forward direction to the gain in the opposite direction out of the back of the antenna. It is expressed in decibels. It is very important in microwave radio backbone systems to have antennas with a good F/B ratio to enable frequency re-use. Ratios as high as 70 dB may be required. When specifying the F/B ratio of an antenna, a wide angle at the back of the dish should be considered and not just the actual value at 180 degrees.

6.2.4 Beamwidth

The beamwidth is an indication of how narrow the main lobe is. The half-power beamwidth is the width of the main lobe at half power intensity (i.e., 3 dB below the boresight gain). The higher the gain of the antenna, the narrower the beamwidth. The reason has to do with the definition of antenna gain. Recall that as the gain is increased in one direction, the sidelobes decrease in another. The beamwidth of the antenna is usually decreased by increasing the size of the reflector. High-gain antennas not only improve the fade margin of a radio link but also result in reduced interference from signals off boresight. One just has to be careful with very high gain antennas that the stability of the towers is sufficient to hold the weight of the large-diameter antennas. Towers must also be rigid enough to avoid a power fade from tower twist. It is not uncommon to have microwave antennas with a beamwidth of less than one degree. With high-gain antennas where the beamwidth is very narrow, angle-of-arrival fading can occur. This causes flat fading due to antenna discrimination. In practice, this limits the useful antenna gain especially on very high frequency links.

6.2.5 Polarization

The polarization of the signal is determined by the hornfeed. Radio links must be set up to transmit and receive on the same polarization. If a signal is received with the opposite polarization, the amount by which the signal is attenuated due to being cross-polarized is referred to as cross-polar discrimination (XPD). Cross-polar operation is often employed on the interference overshoot path as discussed in Chapter 7. Dual polar operation is often used to increase system capacity using a dual polarized hornfeed. This is only possible with solid parabolic antennas. The reflector of a grid antenna is naturally polarized according to the plane of the grid bars and, therefore, will only support one polarization. Dual polar operation usually requires the use of a cross-polar interference canceler (XPIC) to counteract the effect of phase rotation over a hop due to fading mechanisms such as rain.

6.2.6 Radiation Pattern

The pattern set up by an antenna has a three-dimensional aspect. One normally needs to know the shape and amplitude of the various lobes. This is done by plotting the signal around 360 degrees in both planes. In the case of VHF and UHF antennas, this is usually done as a polar plot. For microwave antennas the plot is flattened into a radiation pattern envelopes (RPE) plot, which indicates the envelope of the lobes from −180 degrees to +180 degrees on a linear base. Polar plots and RPEs are compared in Figure 6.2.

6.2.7 VSWR

An antenna presents a complex impedance to the feeder system, which must be attached to it. Since the feeder system also represents a fixed impedance, there can be an impedance mismatch at the antenna connection. Not all the power is thus radiated out the antenna. Some power is reflected back down the feeder. This mismatch is quantified in terms of the voltage standing wave ratio (VSWR). In a real system there will always be some mismatch at both ends. A standing wave is therefore set up in the cable from the reflected waves that are reflected up and down the cable. The cable will attenuate the reflected signal. The reflected signal from the mismatch at the antenna will be re-reflected at the source, however, since it is attenuated by three times the cable attenuation it is not normally a problem. The reflected wave sets up a standing wave with voltage minima and maxima every quarter wavelength. The voltage maxima coincide with points where the incident and reflected waves are in-phase and the minima where they cancel in phase. The VSWR can thus be expressed as

$$VSWR = V_{max}/V_{min} \qquad (6.10)$$

The VSWR value will always be greater than unity and the best VSWR is a value that approaches unity. Practically, a good match will result in a value of around 1.2.

A reflection coefficient (ρ) can be defined that expresses the ratio between the reflected and incident waves

$$\rho = (VSWR - 1)/(VSWR + 1) \qquad (6.11)$$

The most convenient way of expressing this mismatch is the return loss (RL), which is the decibel difference between the power incident on the mismatch and the power reflected from the mismatch. The RL in decibels is expressed in terms of reflection coefficient

$$RL_{dB} = 20 \log(1/\rho) \qquad (6.12)$$

The higher the value of the RL, the better. Typically, this figure should be better than 20 dB for microwave radio systems. To achieve this, individual components should exceed 25 dB.

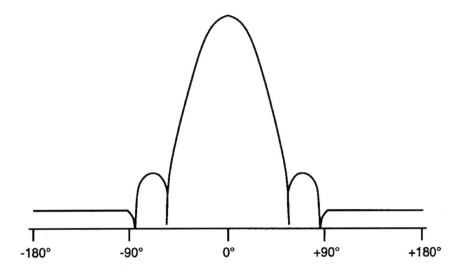

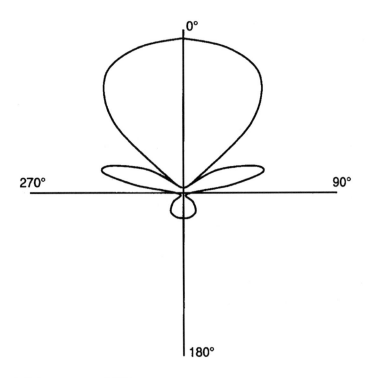

Figure 6.2 Polar pattern and RPE.

6.2.8 Near-Field, Far-Field

Antenna patterns are only fully established at some distance from the antenna. This is the so-called far-field, and geometrical physics can be used to predict field strengths. Concepts such as antenna gain and free-space loss are all defined in the far-field. The signal strength in the near-field is not as easily defined because it has an oscillatory pattern. The far-field distance is defined as

$$\text{Far-field distance} = 2D^2/\lambda \qquad (6.13)$$

where D is the antenna diameter in meters and λ is the wavelength in meters. The effect on antenna gain when antennas are in the near-field is shown in Figure 6.3 [1]. In the first 40% of the near-field the effect is not so drastic, but once this distance is exceeded the response is oscillatory; therefore, it is extremely difficult to predict what the antenna gain is.

This curve is especially useful for back-to-back antenna systems where the short end of the hop may have the antennas quite close together. As an example, assume the antennas were separated by 60m. The gain of an 8-GHz 1.8-m dish is 40.8 dBi in the far-field. The gain of a 3-m dish in the far-field is 45.2 dBi. At 60m the antennas are in the near-field. The normalized near-field factor x is

$$x = R/2D^2/\lambda \qquad (6.14)$$

where R is the distance between the two antennas in meters, D is the antenna diameter, and λ is the wavelength of signal.

Using (6.14) we can calculate that for 1.8-m antennas operating at 8 GHz and separated by 60m, $x = 0.35$. Using Figure 6.3 we can see that the antenna gain reduction is 2 dB. The effective gain of the 1.8-m dishes is 38.8 dBi. Doing the same calculation for the 3-m dishes yields $x = 0.13$. The antenna gain reduction is thus approximately 6 dB. The effective gain of the 3-m dishes is thus 39.2 dBi. It can be seen, therefore, that due to the near-field effect, the effective antenna gain does not increase as the antenna size is increased. The antennas are close coupled, and as the antenna gain increases the coupling factor increases.

6.3 Types of Microwave Antennas

The different antenna types considered here are all based on a parabolic reflector. The hornfeed is always placed at the focal point, so the signal reflected off the reflector element is in phase, as shown by Figure 6.4.

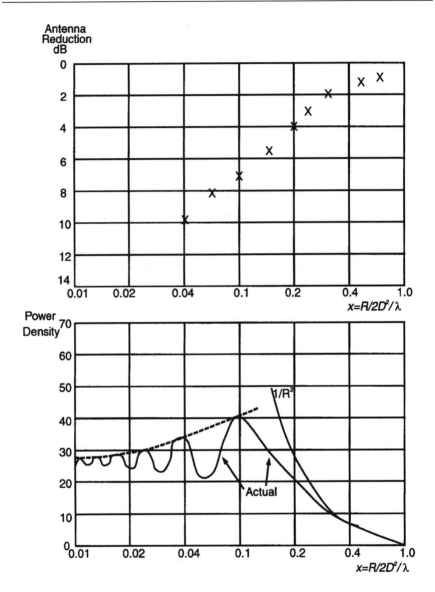

Figure 6.3 Near-field effect on antenna gain.

6.3.1 Grid Antenna

Grid antennas can be used at lower microwave frequencies, below about 2.5 GHz. The advantage of grid antennas is that they have significantly less wind loading on the tower. From an electrical point of view it has the same antenna parameters as a solid dish—exactly the same hornfeed can be used.

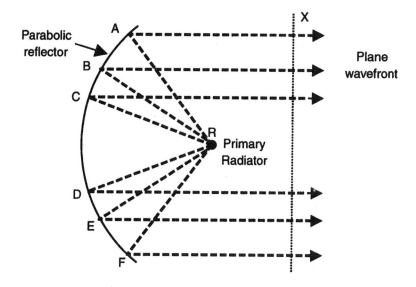

Figure 6.4 Parabolic antenna.

The wavelength is such that the "gaps" between the grids are not required. Electrically there is no difference between a solid reflector and a grid reflector. In practice, the grid reflector has a slightly worse F/B ratio due to diffraction around the grid elements. One limitation of grids is that they cannot support more than one polarization. The reflector rods naturally polarize the signal in the direction that the rods lie. This results in very good cross-polar discrimination.

Grid antennas tend to be significantly cheaper than solid antennas. The hornfeeds are a simpler construction and the amount of material used for the reflector is less. Transportation costs, which make up a significant portion of the cost of an antenna, are also reduced because they can be shipped in a disassembled form.

6.3.2 Solid Parabolic

6.3.2.1 Standard Antenna

Standard parabolic antennas are usually constructed of aluminum. They are manufactured by pressing a sheet of aluminum around a spinning parabola-shaped chuck. The reflectors themselves are not frequency dependent, but the higher the frequency, the greater the surface perfection required. In practice, the reflectors are therefore specified per frequency band. This antenna has standard parameters of gain, F/B ratio, beamwidth, and RL. If one wants an improvement in these parameters, certain changes to the antenna need to be made.

6.3.2.2 Focal Plane Antenna

In order to improve sidelobe suppression and the F/B ratio, the focal plane antenna extends the reflector parabola to the plane of the focus. This means that the aperture area is increased. Rather than increasing the illuminated area, which would result in an increase in gain, the same area is illuminated, reducing the spillover that results in sidelobes and back lobes. In practice, the gain actually reduces due to reduced antenna illumination efficiency. To improve the F/B ratio further, the antenna has special edge geometry. The dish is constructed with a serrated edge. This breaks up the eddy currents, canceling the phase addition components, thus reducing radiation behind the dish. This type of dish offers at least a 10-dB improvement in the F/B ratio compared with a standard antenna, with a very slight reduction in gain (significantly less than 1 dB). The principle of a focal plane antenna is shown in Figure 6.5.

6.3.2.3 High-Performance Antenna

When a very good F/B ratio is required with excellent sidelobe suppression, high-performance, very high performance, and ultra high performance antennas are required. These antennas use a shroud around the edge of the dish to eliminate radiation from the sides and back of the antenna. These shrouds are covered in highly absorbent material that absorbs the microwave radio energy.

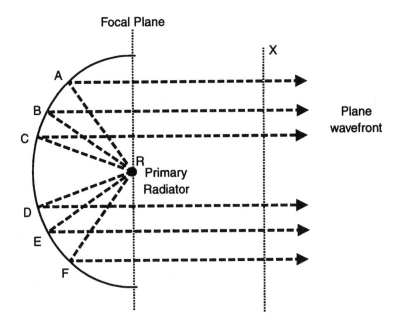

Figure 6.5 Focal plane antennas.

In the case of ultra high performance antennas, the hornfeed itself is also covered with this material to eliminate the scattering off it. This does result in slightly less gain because some of the forward energy is absorbed as well. Special feeds are also employed for this type of dish. The RPEs are specified for each side of the antenna since the hornfeed is not symmetric, and therefore the pattern varies slightly for each side. High-performance antennas are significantly more expensive than standard antennas but are usually mandatory if frequency re-use is required.

6.3.2.4 Slipfit Antennas

For radio systems where the antenna RF unit is mounted outdoors, the antenna is often mounted directly onto the RF unit. This eliminates the need for a waveguide connection, significantly reducing the overall losses. This connection is proprietary because it is developed by a joint design between radio and antenna manufacturers.

6.3.3 Radomes

Special covers for antennas called radomes are available to protect the hornfeed and reduce the wind loading on the tower. These radomes vary in their construction depending on the type of antenna. For standard parabolic dishes the radomes are usually a conical shape constructed out of fiber glass. The radome must be constructed such that its insertion loss is minimized. For shrouded antennas, a fabric radome is usually employed that covers the aperture of the dish. While the unshrouded parabolic dishes can be used with or without radomes, the shrouded construction has such a severe wind loading that the radomes are mandatory. These planar radomes must have a forward-sloped side profile so that water droplets cannot reside on the radome. The high-performance shrouds are thus shaped accordingly. A radome can reduce windloading by a factor of three. Antennas with a diameter of one or two sizes higher, with a radome, can thus present the same wind loading as the smaller antenna without a radome. Radomes are thus highly recommended in any installation.

6.4 Feeder Characteristics

The purpose of the transmission line (feeder) in this context is to transfer the RF signal from the transmit module of the radio equipment to the antenna system in the most efficient manner. For equipment configurations that have the RF unit at the back of the antenna, the feeder is used to carry the baseband

and IF signals plus the power and telemetry signals. There are two main types of transmission lines used in microwave systems: coaxial cables and waveguides.

6.4.1 Coaxial Cable

Coaxial cables are constructed with a metallic inner core with a dielectric material separating the outer metallic conductor, as shown in Figure 6.6.

The cable is covered with a plastic jacket for protection. The dielectric material is usually air or foam, in practice. The cables are fairly robust and therefore easy to install. Different cable jackets are available for indoor and outdoor use. For indoor applications, cables are checked for fire resistance and gaseous emissions. For outdoor applications, the jacket UV resistance is the key factor. The center core of the conductor is invariably copper to ensure high conductivity. The outer cable has a greater surface area and so conductivity is not as critical. Copper-clad aluminum or silver-plated steel wire is also used. The outer conductor often takes the form of braided wire to improve flexibility, however a solid conductor has far superior intermodulation product (IMP) performance. In this case, an annular surface is usually employed to ensure flexibility.

Cable loss is a function of the cross-sectional area; therefore the thicker the cable, the lower the loss. Obviously the disadvantage of thicker cables is the reduced flexibility and increased cost. Cable loss is quoted in decibels per 100m. Air dielectric cables offer a low-loss solution but have the added complexity of pressurization to keep moisture out.

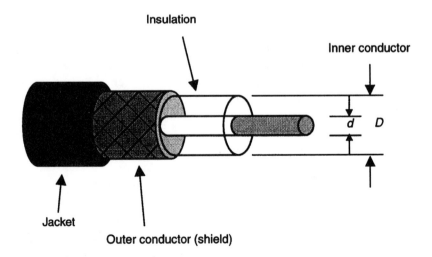

Figure 6.6 Cross-section of coaxial cable.

As the frequency of operation increases, the resistance of the conductor increases, resulting in power loss due to heating. Any alternating current does not have a uniform current density. The current density tends to be greater at the surface of the conductor, which is a phenomenon known as the skin effect. At gigahertz frequencies, this change in resistance can be large. The conductor loss per 100m thus increases as frequency increases.

Another reason that the loss goes up as the frequency increases is that the radiation loss from a conductor increases with frequency to the power of 4 (f^4). The RF energy travels down the conductor as an EM wave with the inner conductor and outer conductor coupling the fields in such a way as to propagate the signal. As the frequency increases, new modes of coupling are introduced that interfere with the signal. This effect coupled with the high-insertion loss limits the usefulness of a coaxial conductor to below about 3 GHz for most applications. Where very short RF cable lengths are required, for example to connect the outdoor RF unit to the antenna in split-unit radio configurations, special coaxial cables can be used in frequency bands as high as 23 GHz.

For applications where there is a split-unit radio configuration, the cable is used to carry baseband, IF, telemetry, and DC power. The attenuation characteristics usually limit the cable length to approximately 150m for a RG-6 type cable. This length can be doubled for an RG-11 type cable. Coaxial cables are available in 50-Ω or 75-Ω versions. The characteristic impedance (Z_0) can be calculated as

$$Z_0 = 138/\sqrt{E} \times \log D/kd \qquad (6.15)$$

where E is the dielectric constant, D is the outer core diameter, d is the inner core diameter, and k the stranding factor.

The usual standard is to use 50Ω for RF applications and 75Ω for IF applications. Traditionally, this is to accommodate the fact that IF cables have longer cable runs. Coaxial cables with 75-Ω characteristic impedance have a lower attenuation value than 50Ω. RF cables tend to carry a higher power signal hence the choice of 50Ω. The output impedance of the radio equipment, the cable connectors, and the cable itself should be matched to ensure a good power transfer and low signal distortion. The connector itself does not have a characteristic impedance unless a dielectric section has been inserted into the connector barrel. One does need to check that the pin dimensions match since traditional connectors have different pin sizes for the different impedance connectors.

6.4.2 Waveguide

Microwave energy can be guided in a metallic tube—called a waveguide—with very low attenuation. The electric and magnetic fields are contained within the guide, and therefore there is no radiation loss. Further, since the dielectric is air, the dielectric losses are negligibly small. A waveguide will only operate between two limiting frequencies, called the cut-off frequency. These frequencies depend on the waveguide geometry compared to the wavelength of operation. The waveguides must be chosen within the frequency band that supports the desired mode of propagation.

Recall that an unguided wave traveling through free space travels as a TEM wavefront. In a guided rectangular waveguide the plane waves are reflected from wall to wall, resulting in a component of the electric or magnetic field being in the direction of the wavefront and another component being parallel to it. It is thus no longer traveling as a TEM wave. This is illustrated in Figure 6.7.

In lossless waveguides the field configurations (or modes) can be classified as either TE (transverse electric) or TM (transverse magnetic) modes. In a rectangular guide the modes are designated as TE_{mn} and TM_{mn}, where m and n denote the integer values for the number of half-wavelengths in the x- and y-directions, respectively, assuming the wave is traveling in the z-direction, as shown in Figure 6.8.

Below a certain critical cut-off frequency, there is no wave propagation in the guide. This cut-off frequency is expressed as

$$f_c = 1/(2\pi\sqrt{(\mu\epsilon)}) \cdot \sqrt{(k_x^2 + k_y^2)} \tag{6.16}$$

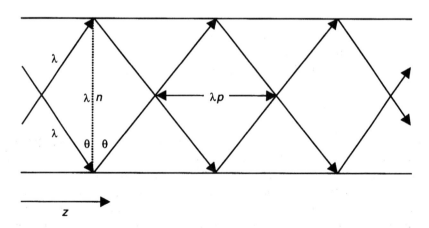

Figure 6.7 Propagation in a waveguide.

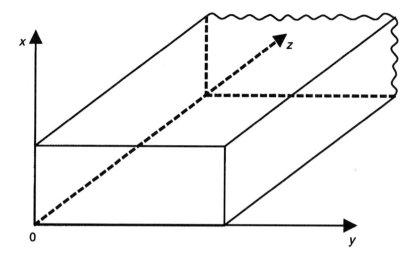

Figure 6.8 Propagation modes in a rectangular waveguide.

where $\sqrt{(k_x^2 + k_y^2)}$ is the cut-off wave number and $\mu\epsilon$ is the dielectric constants of the medium. For a waveguide with height dimensions (a) and width dimensions (b)

$$k_x = m\pi/a \tag{6.17}$$

and

$$k_y = n\pi/b \tag{6.18}$$

Using (6.15) and substituting (6.16) and (6.17) yield the cut-off frequency in a rectangular guide

$$f_c = 1/(2\sqrt{(\mu\epsilon)}) \cdot \sqrt{(m^2/a^2 + n^2/b^2)} \tag{6.19}$$

The cut-off frequency is thus a function of the dimensions of the guide. The physical size determines the mode of propagation. In a rectangular guide the width is usually twice the height, hence $2a = b$.

The mode with the lowest cutover frequency is called the dominant mode. In a rectangular mode this is the TE_{01} mode [1]. If we recall that $1/(\sqrt{(\mu\epsilon)}$ is equal to the velocity of propagation and that the dielectric constant of air can be assumed to be equal to unity, then the cut-off frequency for a rectangular guide operating in the TE_{01} mode can be expressed as

$$f_c = c/(2\sqrt{(1/b^2)}) \qquad (6.20)$$

$$= c/2b$$

where c is the speed of light (3×10^8 m/s). For TE_{10}, $f_c = c/2a$.

It is possible to have more than one mode in a waveguide, however usually only the dominant mode will propagate and the other modes decay very quickly.

Another important concept in waveguide theory is that the wavelength inside the guide differs from that outside. The wavelength inside the guide for TE_{mn} modes can be expressed as

$$\lambda_g = \lambda/\sqrt{(1 - (f_c/f)^2)} \qquad (6.21)$$

Due to the complex field configurations and strong dependence on the shape of the waveguide, installations must be conducted with utmost care. Waveguides have a limitation on bending in both planes. The waveguide should not be twisted because this results in unquantifiable group delay distortion.

The most common waveguide used in a microwave radio installation is the elliptical waveguide. This has corrugated copper walls with a plastic sheath for protection. The corrugations result in a strong cable with limited bending ability. The limitation on bending is specified in terms of a bending radius in the E-plane and the H-plane. A much smaller bending radius is allowed in the E-plane; therefore, one should utilize this when planning a waveguide installation. Although a maximum twist allowance is specified, twists should be avoided when planning an installation. A change in plane from E to H can easily be achieved within a few meters without twisting the waveguide by bending the waveguide within the specified bending radius of each E- and H-plane, respectively. The effective usable length of waveguide is determined by the loss of waveguide. The waveguide loss is specified in decibels per 100m and increases significantly as frequency increases. Above 10 GHz, the loss becomes excessive and radio manufacturers often offer the choice of a baseband or IF connection to an outdoor RF unit mounted on the tower to avoid long lossy waveguide runs.

6.5 Antenna System Accessories

6.5.1 Pressurizers

In order to avoid the ingress of moisture into an air dielectric feeder, it should be pressurized. Daily temperature variations result in a variable volume of air inside the feeder. During nocturnal cooling, warm moist air can penetrate the feeder, resulting in moisture on the feeder walls that can result in unwanted oxidation. Dry air or an inert gas such as nitrogen is used with a pressure of about 10 mbar to prevent moisture ingress. A pressure window needs to be inserted between the pressured and nonpressurized sections of the feeder, which is usually located at the radio flange after the branching. The pressurizer unit needs to be sized according to the total length of feeder that needs to be pressurized. Dry air pressurizers are often called dehydrators. Conventional types force the moist air through a desiccant, which absorbs the moisture. More recent types use a membrane to separate the moisture.

6.5.2 Plumbing

Waveguide connections onto the radio and antenna can be made easy using a combination of rigid straight sections, twists, and bends. These need to be carefully planned by measuring the height of the waveguide entry relative to the waveguide port on the equipment and by taking into account the polarization of the antenna.

6.5.3 Earth Kits

It is important not to allow induced surges into the equipment room via the feeder cable. The cable should thus be properly earthed. Earthing kits are available that provide good contact between the waveguide itself and the Earth strap. When installing these it is important to seal the unit so that moisture is not allowed into the connection. The shortest and straightest path should be sought between the earthing kit and the Earth point. For coaxial cables, inline arrestors are available that use gas arrestor tubes to absorb the surges. One problem with these tubes is that often they can be discharged without the user being aware that the tube has blown. Regular testing is thus required.

6.5.4 Cable Clamps

Cables on a tower are subject to severe stresses and strains due to vibration from wind forces. It is very important that they are securely clamped to the waveguide ladder. In the case of waveguide it is also important to support the weight of the cable, otherwise the corrugations of the feeder at the base of the tower may start to bunch. It is normally recommended that the clamps are placed no more than 1-m apart. The clamp must also not cut into or distort the feeder. Plastic ties get brittle very quickly due to ultraviolet (UV) radiation from the sun. A UV resistant and corrosion-free material such as stainless steel is thus normally required.

6.6 Installation Practices

Antennas can weigh half a ton and are subject to severe wind forces. They are also intended to be operational for twenty or more years. It is very important, therefore, that they are installed properly and that the necessary measures are taken to ensure long-term survivability. Galvanized bolts should be used for all fittings. Stainless steel is a very good corrosion-free material, but it tends to creep and therefore should be used with caution on threaded items. The bolts must be long enough to protrude through the nut with enough space for the necessary spring washers and lock nuts.

Large antennas usually have side struts to stop the antenna from twisting on the tower. These struts must be secured to a strong rigid member on the tower. Some cross-bracing members on the tower are inadequate to support these struts, so a careful choice of bracing members is essential.

For safety reasons, panning arms, side struts, and feeder cables should not obstruct the walkways. Earth straps should be installed at the top and bottom of the feeder on the tower at the extremes of the vertical run. In addition, if the horizontal run from the tower to the building exceeds about 5m, an additional Earth kit should be installed at the waveguide entry.

Part of the installation and commissioning of a radio system is to achieve the calculated receive level. This is done by aligning (panning) the antennas. As discussed earlier, microwave antennas are directional and the maximum system gain is achieved by ensuring that the main signal out of the boresight of the antenna A is received into the boresight of antenna B. One of the most common errors when commissioning a radio system is to have the antennas panned onto a sidelobe. This can be avoided by panning for maximum strength over a wide angle. Once a maximum signal is achieved, further panning should result in an initial reduction of the signal followed by a gradual increase to a new peak at a lower level. This confirms that the original peak was the main lobe, and therefore the antenna should be repanned to this position. Provided that accurate measurements are taken, the antennas should be panned to achieve within 3 dB of the calculated receiver level. When this is achieved it is a good indicator to the system planner that the system design is sound.

All bolts and fixing materials must be protected against corrosion. Self-vulcanizing sprays are available to seal off any fittings. The feeder to antenna connection and earthing strap connection should also be protected against the elements by sealing it with self-vulcanizing tape covered with a UV-resistant scotch tape.

The labeling of all feeders is a good installation practice. Labels should be made with a corrosion-free material such as copper and firmly secured on the feeder with a soft, yet strong strap to avoid damage to the feeder.

Future expansion at the site must be taken into account when planning the installation. Feeders should never be installed diagonally across the gantry and waveguide ladder. Due to the fact that antennas are often placed in locations where specially qualified riggers are required for installation, it is important that the initial installation is thoroughly inspected before the site is handed over for commercial operation. This will reduce any future outage problems during the operational life of the system.

References

[1]   Liao, S., *Microwave Devices and Circuits*, Englewood Cliffs, NJ: Prentice-Hall, 1980.

7

Frequency Planning

Planning a radio system involves ensuring that the radio system will carry traffic with the desired quality level. A lot of discussion goes into understanding fading mechanisms to ensure that the radio system is designed to overcome these unwanted effects. One effect that has been ignored until now is interference. Frequency planning is such an important aspect of radio link design that a whole chapter has been devoted to it.

7.1 What is Interference?

Interference is any unwanted signal that would present itself to the receiver section of a radio for demodulation. It can be a delayed copy of the radio links' own signal, an adjacent channel's signal traveling over the same link, or a signal from another radio link or RF source.

7.2 Causes of Interference

The cause of interference can be divided into two categories: internal and external.

7.2.1 Internal Causes

Internal causes are those causes that relate to the equipment at the site itself. This includes radio equipment parameters such as the transmit and receive local oscillators (LOs) and filter selectivities. It also includes internal aspects of the system design such as reflections from the antenna/feeder system, the transmit/receive (T/R) spacing, the F/B ratio of the antennas at a repeater station, and cochannel and adjacent channel interference from the system itself. These factors are all basically under the system designer's control. Good equipment with stable oscillators and steep filter responses is required. In addition, international channel plans should be chosen that incorporate stringent interference limitations. Finally, the antennas must be chosen to meet the interference requirements.

7.2.2 External Causes

External causes result from sources that are seldom under the designers control. These include interference from other systems that are already installed and interference from other services such as satellites. It also includes interference from a distant site that forms part of the same route—this aspect can be controlled by the system designer.

7.3 Types of Interference

There are two types of interfering signals: one with a variable C/I, the other with a constant C/I.

7.3.1 Variable C/I

In this case the interfering signal is constant but the carrier level varies due to fading over the path. It is assumed that the fading that the carrier experiences is independent from the fading experienced by the interfering signal. This results in the C/I vary with fading. The effects are thus analyzed at the threshold of the receiver (high BER).

7.3.2 Constant C/I

In this case both the carrier and the interfering signal are affected by the same amount of fading. This is the case when the signals travel over the same path. The absolute wanted signal and interfering signal levels may change, but the ratio between them remains the same. These effects are analyzed for a strong receive signal where thermal noise is not a problem and intersymbol interference is the main concern.

7.4 Effects of Interference

In an analog system the interference effect is totally different from a digital system.

7.4.1 Effect on Analog Receivers

In analog systems the interference increases the idle and baseband noise, which has a direct effect on quality. It also leads to a build-up of IMP, which further degrades the signal quality. These products interfere with the carrier frequency and its sidebands. An interfering signal's carrier frequency, received within the victim receiver's band, can have a much stronger signal level than the victim's sidebands. This results in what is called carrier beat interference. The second type of interference is caused by the adjacent channel's sidebands, which beat with the victim's sidebands resulting in sideband beat interference. These analog interference effects are shown in Figure 7.1.

The interference will also add to the thermal noise floor of the receiver. Background thermal noise (P_n) can be quantified as

$$P_n = KTB \tag{7.1}$$

where K is Boltzman's constant (1.38×10^{-23} J/K), T is the receiver temperature (in Kelvins), and B is the bandwidth of the receiver in Hertz.

The receiver threshold P_T can be expressed in dBm

$$P_T = S/N_{BB} + F_{dB} + N_i + KTB \tag{7.2}$$

where S/N_{BB} is the S/N ratio at the demodulator input, F_{dB} is the receiver front-end noise figure, and N_i is the noise from interference. This results in an increase in the baseband idle noise and degrades the S/N ratio at the RF squelch point, thus decreasing the fade margin. This effect is negligible in comparison with the previous effects in analog systems.

The effect of increased noise due to thermal noise and intermodulation is shown in Figure 7.2. It is interesting to note that the crossover point is the desired operating point, which is why most analog systems were designed to meet a specific fade margin (usually 40 dB) irrespective of the link conditions.

In summary, it can be said that the main effect on analog systems is due to increased noise and IMPs in the unfaded condition.

7.4.2 Effect on Digital Receivers

In an unfaded condition, digital receivers are very robust against interference mechanisms. Unlike analog systems, however, the main interference problem

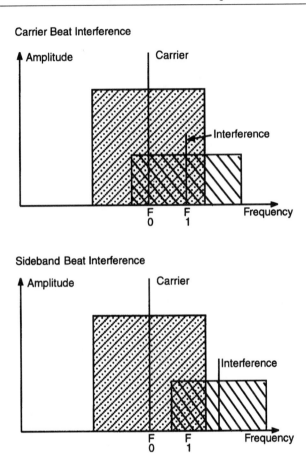

Figure 7.1 Carrier beat and sideband beat interference.

occurs in a faded condition where the signal levels approach the receiver threshold values. Signal levels way below the threshold limit of the receiver can cause problems to the demodulation process. The interference effect is thus not in terms of its absolute signal amplitude but in terms of the ratio between the wanted (carrier) signal and the unwanted (interference) signal, expressed as C/I. If the figure is expressed relative to the threshold level it is called threshold-to-interference, or T/I. The C/I ratio is also sometimes called wanted-to-unwanted, denoted W/U. The two terms can be used interchangeably.

7.4.2.1 Cochannel Interference

In digital systems, due to the threshold effect of digital receivers as illustrated in Figure 7.3, the low-level interference has little or no effect on signal quality

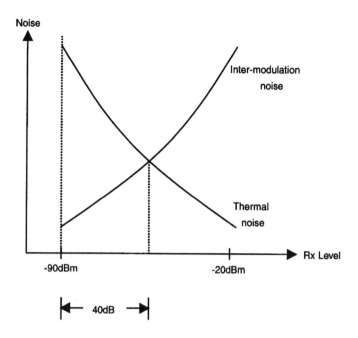

Figure 7.2 Noise in an analog system.

in the unfaded condition. For small variations around the nominal receive level, the effect on BER is negligible. It is only when one approaches the knee of the threshold area that receive level variations have a dramatic effect on quality. It is interesting to note that due to the threshold effect it is not essential to operate at a specific fade margin, for example, 40 dB. One can calculate the actual fade margin required for the specific link provided that a minimum fade margin is maintained to keep the signal off the threshold portion of the curve.

Recall from (7.2) that interference N_i adds to the thermal noise floor of the receiver. If the threshold level without interference was −100 dBm and the interfering signal had an amplitude (N_i) of −100 dBm, the threshold would be degraded by 3 dB. This means that a fade margin of 40 dB would be degraded to only 37 dB, which has a significant effect on the overall performance of the link.

In a digital system there is a certain minimum C/I ratio (C/I$_{min}$) above which the BER is constant and below which the performance quickly becomes unacceptable. This depends very much on the modulation scheme: A simple 4 PSK system requires only 15 dB whereas a 128 QAM system requires at least 30 dB. The digital signal must maintain this ratio even in a faded condition. This means that the C/I must be greater than C/I$_{min}$ plus the minimum fade margin calculated to meet the performance objective, hence

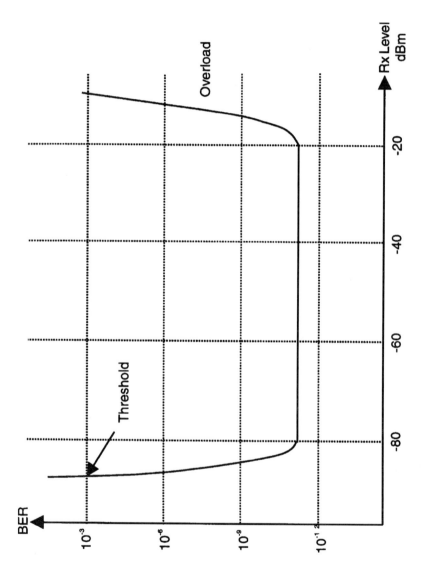

Figure 7.3 Threshold effect of digital systems.

$$C/I = C/I_{min} + FM_{min} \qquad (7.3)$$

In practice, the interference situation at a nodal site improves if the receive levels are of equal magnitude. The receive levels are affected by the antenna sizes used over the respective hops. This requirement must be balanced with the performance objectives that dictate specific antenna sizes in order to meet a required fade margin. This is illustrated in a tutorial example later in this chapter.

Equipment manufacturers will usually produce a series of curves that show the threshold degradations for various C/I ratios. An example is shown in Figure 7.4.

The threshold degradation needs to be included in the overall fading predictions. The overall fade margin is a combination of flat fading, selective fading, and interference. The fade margin calculated in the previous chapters needs to be reduced by the value of the threshold degradation to work out the overall outage.

7.4.2.2 Adjacent Channel Interference

In order to simplify the analysis, the filter discrimination, called the net filter discrimination (NFP, or interference reduction factor (IRF)) is used to convert the adjacent channel interference into an equivalent cochannel value. This value is a function of the selectivity of the receiver itself and, therefore, must be obtained from the manufacturer's specifications for each radio type. An example is shown in Figure 7.5. Using the concept of NFD, the same analysis

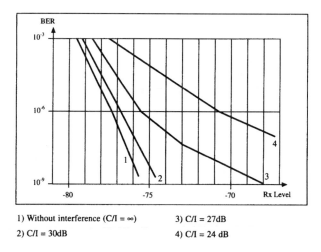

1) Without interference (C/I = ∞)   3) C/I = 27dB
2) C/I = 30dB   4) C/I = 24 dB

Figure 7.4 Example of equipment C/I curve.

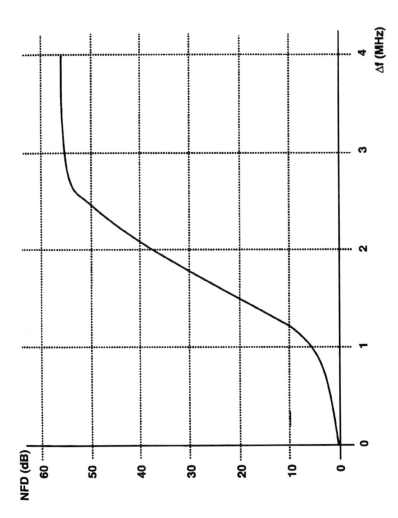

Figure 7.5 Typical equipment NFD curve.

can be used for cochannel and adjacent channel interference signals. The equation can be expressed as

$$C/I \text{ (cochannel)} = C/I \text{ (adjacent)} + NFD \qquad (7.4)$$

7.5 Interference Conditions

Channel plans for analog systems have been in place for many years. With new digital systems and, specifically, SDH radio systems being introduced, the challenge is to fit the new services into the same spectrum. This has been done using more complex modulation schemes and with dual-polar operation on the same channel. This results in three interference conditions: digital to digital, analog to digital, and digital to analog.

7.5.1 Digital to Digital

In a digital system the effect of an interfering signal can be analyzed using the C/I curves supplied by the radio manufacturers. These curves assume the interfering cochannel signal is of the same type as its own wanted (carrier) signal, using the same modulation scheme. A family of curves is usually supplied starting with a C/I ratio of infinity (no interference) up to a level where the system becomes unusable (C/I_{min}). The curves are usually only shown over the threshold region, but they can be supplied over the entire range of receive levels including strong (overload), receive levels. The adjacent channel interference can be calculated using the concept of NFD. The carrier and interfering signal levels must be calculated taking into account the antenna discrimination for off-boresight signals, cross-polarization, and any diffraction losses. In other words, it is the actual signal levels received at the antenna port.

7.5.2 Analog to Digital

Analog systems tend to have a narrower bandwidth than the equivalent digital systems; therefore, interference from an analog into a digital system is less severe than vice versa. The cochannel interference situation can be considered similar to the digital-to-digital analysis providing one worsens the results slightly (approx. 1 dB) to take account of the fact that the interfering analog FM signal spectrum can be considered unmodulated.

The adjacent channel signal results are usually improved on the digital-to-digital case due to the analog signal's narrower bandwidth. The receiver selectivity attenuates the concentrated analog signal more than the wideband digital system.

7.5.3　Digital to Analog

This is the most serious interference situation. In other words, the digital systems being introduced have a much worse effect on the analog systems in place than those analog systems will have on the new digital systems. Fortunately, in most cases the new digital systems are actually replacing the analog ones, and therefore the performance degradations can be tolerated during the cutover period. The performance calculations will not be covered in detail because most analog systems are now obsolete. As mentioned earlier, the interference results in increased noise. This is normally referenced to the worst FDM channel, which is the highest baseband frequency. The interference calculation must take into account the main digital interference, the secondary noise from the PCM spectrum beating with the FDM carrier, and the more serious IMP noise caused by IM products resulting from the two sets of spectrum.

7.6　Frequency Channel Planning

Recommendations specifying the individual radio channel are made by the Radiocommunications sector of the ITU. These ITU-R Recommendations specify the center frequency of the band, the T/R spacing, the adjacent channel spacing (copolar and cross-polar), and the number of channels.

7.6.1　Basic ITU Arrangements

The basic frequency arrangement set by the ITU-R is shown in Figure 7.6.

The center frequency (f_0) is the midband frequency around which the channels are arranged. A number of channels with a specific channel spacing are identified across the frequency band of that specific plan.

Sometimes the channels are spaced with alternate polarizations. One therefore needs to be careful when specifying channel spacing requirements whether it is the copolar or cross-polar spacing that is being referenced. A safety gap (center gap) is also specified between the top frequency of the transmit-low frequencies and the bottom frequency of the transmit-high frequencies.

The ITU-R plans are usually provided with the basic configuration showing the center frequency f_0, the channel spacing, the T/R spacing, and the overall frequency band. A formula is then provided from which one can calculate the actual frequencies themselves. The actual frequencies for an international connection [1] in the frequency band 7425 MHz to 7725 MHz using f_0 of 7575 MHz is calculated as follows:

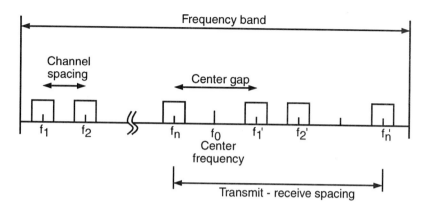

Figure 7.6 ITU-R frequency plan arrangement diagram.

$$fn = f_0 - 154 + 7n \quad \text{and} \quad fn' = f_0 + 7 + 7n \qquad (7.a)$$

Hence,

$$f1 = 7575 \text{ MHz} - 154 + 7 \times 1 = 7428 \text{ MHz}$$
$$f2 = 7575 \text{ MHz} - 154 + 7 \times 2 = 7435 \text{ MHz} \qquad (7.b)$$
$$f1' = 7575 \text{ MHz} + 7 + 7 \times 1 = 7589 \text{ MHz}$$

It can be seen that the channel spacing is $(7435 - 7428) = 7$ MHz and the T/R spacing is $(7589 - 7428) = 161$ MHz. Radio equipment's duplexers will usually only support one T/R spacing. Therefore, the planner should specify the T/R spacing of the frequency plan when ordering the equipment.

7.6.2 A and B Sites (High/Low Arrangements)

For radio links one is always working with pairs of channels. A signal is transmitted from Site a to Site b with a certain transmit frequency and Site a's receive frequency is the transmit frequency from Site b. These are termed go and return channels. The go channels transmit in the lower half of the plan and are sometimes referred to as "transmit low" and are designated as f_n, where n is the channel number. The return frequencies transmit in the higher portion of the band, are sometimes referred to as transmit high, and are designated as f_n'.

This is a very important consideration when doing the frequency coordination at a site. For a specific frequency band all the links must transmit either high or low, otherwise one can end up with a situation where a strong transmit signal leaks into the adjacent link's receiver that may not be able to adequately filter it. Transmit-low sites are called A sites and transmit high sites are called B sites. It is possible to have one site as an A site for one frequency band and a B site for another frequency band if the frequency separation is sufficient.

One does need to be very careful with adjacent frequency bands because the return channels from the lower band are adjacent to the go channels for the higher band. It is thus imperative that if the site is a B site for the lower band it must be an A site for the higher band. This is illustrated in the example shown in Figure 7.7. Radio routes that link up in the form of rings must be carefully planned to avoid an A/B clash. In this case it may be necessary to use another frequency band in the loop to break the frequency chain. The main problem with an A/B clash is that the transmitter at the site can interfere with its own receiver. One needs to maximize the frequency separation between transmit and receive. To do this, channels at this problem site should be chosen at the extremes of the band. For example, with an eight-channel band plan, if in one direction one is using channel 1, try to use channel 8 in the other direction. This way, the T/R spacing is only the "center gap distance" short of the usual T/R spacing ensured by the band plan. To illustrate these principles consider Figure 7.7 that represents a five-hop network. Assume that the hop frequencies for channels 1 to 3 are as indicated in Table 7.3.

Given that site a has a transmit frequency of 7750 fixes it as an A (transmit low) site. Site b is thus a B (transmit high) site. It will transmit

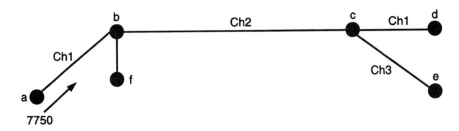

Figure 7.7 High/low frequency illustration.

Table 7.3
Available Frequencies for a High-Low Example

Channel	Low Frequency (MHz)	High Frequency (MHz)
1	7750	8050
2	7780	8080
3	7810	8110

8050 MHz to site a and 8080 MHz to site c. Site c must be an A site (transmit low), which fixes sites d and e as B sites (transmit high). Site c will thus transmit 7780 MHz to b, 7750 MHz to d, and 7810 MHz to e.

Now assume that a new site f is added with frequencies 7400 MHz and 7700 MHz. In order to allocate the frequency, we need to decide which end is A and which is B. Site b is already a B site for the previous band plan, which means it receives on 7750 MHz. If we kept the A/B allocation the same for this plan, the transmit frequency from site b would be 7700 ("transmit high"), which is only 50 MHz away from the other link's receiver. The duplexer isolation can probably not cope with this; therefore, it is better to reverse the allocations. This is always the case with frequencies at the limits of their respective bands, that is, when two plans are adjacent to one another. The solution is thus to allocate site b as an A site (f_{TX} = 7400) and Site f as a B site (f_{TX} = 7700). It would also be unwise to close the loop between e and d on the same plan because it would cause an A/B clash. It is preferable to implement one of the links in a different band. For example, if the distance between e and d was less than 20 km it could probably be linked on a frequency above 10 GHz. If there are no alternative frequency bands and no other way of doing the route, then frequencies at the edge of the band must be chosen; that is, with Link ce on channel 3 and Link cd on channel 1, channel 8 should be chosen for Link ed, assuming an 8-frequency channel plan. The T/R isolation between channels 3 and 8 must then be checked relative to the duplexer isolation. If it is insufficient, a channel filter could be investigated; if this is not practical the link cannot be installed. Allocating the high/low ends is always the first step in doing any channel allocations during the frequency planning phase of the link design.

7.6.3 Alternate Polarization

In fully developed routes, alternate polarizations are used with a dual-polar antenna. A typical configuration showing the frequency plan and antenna connection diagram is shown in Figure 7.8.

7.6.4 Frequency Diversity Allocations

In some cases, to meet the specified performance objectives in a route, a protection channel is required. The concept is that when fading activity occurs over the hop, the effect on one frequency is different from an adjacent frequency due to the electrical path length differential between the two. If during fading the traffic is switched, using a hitless switch, to the standby channel, traffic loss can be avoided. The greater the distance between the two frequencies, the

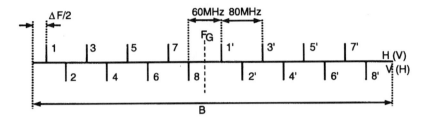

ΔF :   RF Channel spacing (Co-polar)

B   :   RF Bandwidth

F_G :   Center Gap

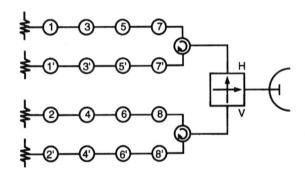

Figure 7.8 Typical dual-polar plan and antenna connection.

less the correlation and the better the performance improvement against Raleigh fading. The partial uncorrelation between two channels in frequency diversity mode is shown in Figure 7.9.

For selective fading the situation is the opposite. The probability of notch fading occurring simultaneously in frequency channels separated by a large spectral distance is higher than if they were spaced close together. With modern radio systems, however, the notch distortion is equalized with fast-responding equalizers so that the dominant fading outage is still Raleigh fading. Typically only 10% of the outage can be attributed to selective (notch) fading. When allocating frequencies, one should therefore aim for maximum separation. In a 1 + 1 system, one needs to consider that each channel used must have an equal performance; therefore, the separation is never more than half the number of channels. For example, if there are eight channels in a band and the entire band was to be used, the diversity would be allocated as 1, 5 and 2, 6, for example. If $n + 1$ diversity is used, it implies that the n channels have only one standby channel. It is impossible for each channel to have equal perfor-

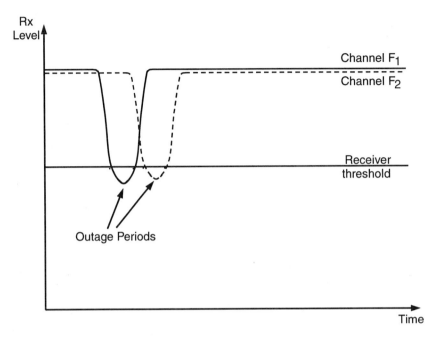

Figure 7.9 Frequency diversity improvement against fading.

mance; therefore, if there are eight channels available, the protection channel can be arbitrarily chosen as channel 8. With this configuration one should put the priority traffic on channel 1 to ensure maximum separation between traffic and standby channels.

7.6.5 Interleaving of Channels

In some cases where the channel spacing exceeds the width of the modulated carrier, additional channels can be interleaved in the "gaps," as shown in Figure 7.10.

The intended application for interleaved channels is at nodal sites. It is important to realize that it is not possible to use interleaved frequencies simultaneously with the main plan over a fully developed route. Adjacent channel interference is limited at nodal sites because the signals are not parallel but are discriminated by the antenna patterns.

7.6.6 Spectral Efficiency

The bandwidth of radio equipment often has to be adapted to fit within the specified frequency plans. Complex modulation schemes are used to reduce

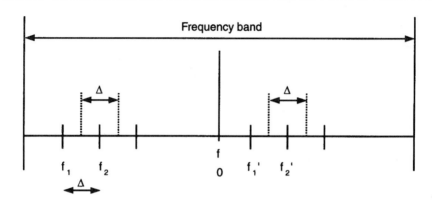

Figure 7.10 Interleaved channels.

the symbol bandwidth so that the channel spacing conforms to the ITU plan. The equipment needs to meet the required spectral efficiency for each plan. Spectral efficiency (E) is the ratio between the information bit rate and the RF bandwidth. It is defined as

$$E = \text{Number of RF channels} \times \text{Maximum bit rate/RF half-band}$$
$$(7.5)$$

For example, a 500-MHz-wide frequency band with eight pairs of RF channels (go and return) with equipment operating at 140 Mbit/s will require a spectral efficiency of

$$E = 8 \times 140/(500/2) \qquad (7.6)$$
$$= 4.48 \text{ bits/s/Hz}$$

This will influence the type of modulation scheme used.

One can calculate the achievable efficiency of different modulation schemes as follows: Assume an ideal situation where the baseband filters are limited to the Nyquist bandwidth. In other words, the filter bandwidth is one-half the symbol rate (T_s). If we assume that B is the number of bits per RF symbol, then the packing ratio or bits per Hertz of RF bandwidth capacity (C) is equal to B. In an M-ary QAM system, B can be expressed as

$$B = 2 \log_2 M \qquad (7.7)$$

This means that for an ideal filter case, $C = 2 \log_2 M$. For a 4 PSK system, $C = 2$. For a 16 QAM system, $C = 4$. For a 32 QAM system, $C = 5$.

In the example discussed at the beginning of this section, it can be seen from the result of (7.6) that a minimum of 32 QAM would be required for this system to fit into this band plan.

7.7 Frequency Re-Use

Frequency re-use refers to a situation where the same frequency pair is re-used in a route. In the interests of good spectrum management, when doing frequency planning one should start from the assumption that the frequency can be re-used.

7.7.1 Two-Frequency (One-Pair) Plan

The most efficient re-use plan is one in which a single pair of frequencies is used throughout a route. One needs to consider the interference from two perspectives: the interference at the repeater (nodal) site and the problem at sites further down the route (overshoot).

7.7.1.1 Nodal Sites

Consider a case where at site A two links on the same frequency are transmitting in two opposite directions. Let the left-hand link be called Link x and the right hand link be called Link y. The situation is as shown in Figure 7.11.

Site A is transmitting frequency f_1 in both directions. Site X is therefore transmitting on frequency f_1'. At site A, the antenna (X) panned for Link x will receive the signal on frequency f_1'. The antenna (Y) panned for Link y will also receive this signal from x, but attenuated by the F/B ratio of the Y antenna. To re-use the same frequency f_1 on Link y, the minimum carrier (signal f_1' from Link y) to interference (signal f_1' from Link x attenuated by F/B ratio of Y) ratio must be maintained into receiver Y. If the interference signal is too high, the F/B ratio of antenna Y will have to be increased. Changing the polarization of Link y will not help because the polarization discrimination

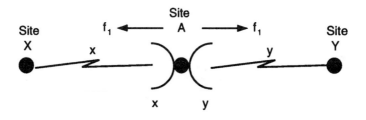

Figure 7.11 Nodal interference.

around the back of the antenna is negligible. Increasing the wanted signal into Y from Link y may help but may cause problems for Link x. The best interference scenario is to have the two signal levels at Site A balanced, if the performance calculations allow it. In summary, the only real way to achieve frequency re-use is to use high-performance antennas with a good F/B ratio.

7.7.1.2 Overshoot

Re-using the same frequency not only has implications at a nodal site, it also needs to be considered from the overshoot point of view. Consider the diagram shown in Figure 7.12.

Assume we have four sites with three links on the same frequency sets. At Site 1, Link x will transmit frequency f_1. This will not be received at Site 3 since this is also a receive high (A) site. The overshoot path is thus into Site 4 on Link z. The interference path is attenuated by the two antenna discriminations. The interference transmit signal is reduced by the antenna discrimination at angle α (Link x) and the receive antenna is reduced by the antenna discrimination at angle θ (Link z).

The receive signal is attenuated over the hop by the FSL, which can be expressed as

$$FSL = 92.4 \text{ dB} + 20 \log f \text{(GHz)} + 20 \log d \text{ (km)} \qquad (7.8)$$

Distance in itself does not afford the hop much protection since a doubling of distance only results in a 6-dB loss in signal level. In other words, for a 50-km-long hop, an interfering signal 100 km away is only attenuated by 6 dB due to the path length. One 400 km away is attenuated by 18 dB. Considering that fades of greater than 40 dB can occur due to multipath fading, an interference signal attenuated by only 18 dB is not negligible.

Fortunately, the Earth's topology often blocks interfering signals that emanate from transmitters that far away. The additional loss by any potential terrain obstruction from the intervening terrain can be included. This must

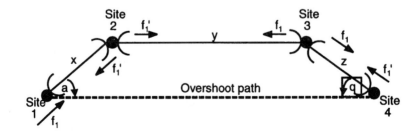

Figure 7.12 Overshoot interference.

be analyzed assuming a so-called flat Earth. In other words, the worst case interference scenario is for a high k condition. The diffraction loss should be calculated for k = infinity. The implication is that even if the obstruction is physically blocking the path from an LOS point of view, it may have no effect from an interference point of view since the radio beam is bent over the obstruction in a high k condition. A signal propagated over a flat Earth with no hills would not be blocked by the Earth bulge at k = infinity because the signal travels parallel to the Earth at this k-factor. Despite this, there are cases where, even in a high k condition, the intervening terrain adds an obstruction (diffraction) loss to the path. This loss should therefore be added to the FSL to calculate the level of the interference signal.

If the interference signal is problematic the only practical way to improve the situation, assuming antennas with good sidelobe suppression are already in use, is to change the polarization of one of the two paths, that is, either Link x or Link z. This way the interfering signal experiences the antenna's cross-polar discrimination, which is significant. For this reason, overreach problems are solved by alternating the polarization every two hops. This plan is shown in Figure 7.13.

7.7.2 Four-Frequency (Two-Frequency Pair) Plan

If the cross-polar discrimination of antennas is insufficient to solve the overshoot problem using a single frequency pair, a second set of frequencies is required. In this case, the frequency and polarization should be alternated on every second hop. This results in the overshoot path to the third hop being a different frequency, and the opposite polarization ensures that adjacent channel interference is minimized. This is illustrated in Figure 7.14.

If the second pair of frequencies is required to overcome antennas with insufficient F/B ratio, the frequencies should be alternated every hop and the polarization alternated every third hop. The common polarization every second node is not a problem as far as the F/B ratio is concerned since there is virtually no polarization discrimination at the back of antennas. The polarization shift

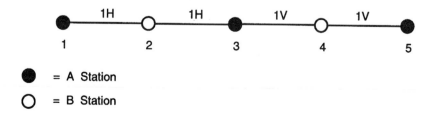

Figure 7.13 Two-frequency plan: cross-polar.

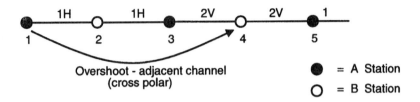

Figure 7.14 Four-frequency plan with alternated polarization.

at the third hop is required for overshoot protection. This is illustrated in Figure 7.15.

With this four-frequency plan, high-performance antennas will not normally be required on the alternate sites where two frequency pairs are being used.

7.7.3 Six-Frequency (Three-Pair) Plan

If both F/B and overshoot cannot be solved with the previous plans, then an additional frequency pair will be required. These will be allocated as frequency pairs 1, 2, and 3 with one polarization, then 1, 2, and 3 with the alternate polarization, and so on.

7.7.4 Practical Interference Example

Interference analysis principles are best illustrated by worked examples. The tutorials in this section are included to demonstrate the basic principles of frequency re-use.

7.7.4.1 Nodal Interference Example

Consider a typical node in a radio route, as shown by Figure 7.16.

Tutorial Problem

Assume that the nominal receive (Rx) level at Y from transmitter X {TX(X)} is designated as C_{YX}. C_{YX} is given as −40 dBm. Assume that the nominal Rx

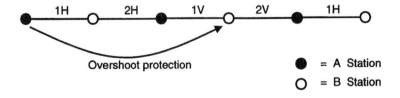

Figure 7.15 Four-frequency plan with alternated frequencies.

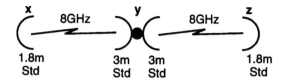

Figure 7.16 Nodal interference example.

level at Y from TX(Z) is C_{Yz} = −45 dBm. Assume that the antennas at Y are both 3-m standard (STD) dishes (8 GHz) with a F/B ratio of 52 dB and antenna gain (A_e) of 45 dBi. Assume the antennas at X and Z are both 1.8m STD (8 GHz) with a front to back ratio (F/B) of 48 dB and antenna gain (A_e) of 40 dBi. Assume the fade margin required to meet the performance objectives is 40 dB and the minimum C/I ratio required by the modem (C/I$_{min}$) is 15 dB. The objective of this tutorial is to determine if the same frequency pair can be re-used, in other words, determine if the frequency used on hop XY can be re-used on hop YZ.

Tutorial Worked Solution

First consider the receive level at Y from the transmitter at X. The carrier level C_{YX} is given as C_{YX} (unfaded) = −40 dBm. Thus, the faded level can be calculated using

$$C_{YX} \text{ (faded)} = \text{Nominal receive level} - \text{fade margin}$$
$$= -40 \text{ dBm} - 40 \text{ dB} \qquad (7.9)$$
$$= -80 \text{ dBm}$$

The interference signal at Y from Z (I_{YZ}) can be derived

$$I_{YZ} = C_{YZ} - F/B \text{ (Y antenna to X)}$$
$$= -45 \text{ dBm} - 52 \text{ dB} \qquad (7.10)$$
$$= -97 \text{ dBm}$$

Using the results from (7.9) and (7.10), the carrier (C_{YX}) to interference (I_{YZ}) ratio is shown by

$$C/I \text{ (faded)} = C_{YX} \text{ (faded)}/I_{YZ}$$
$$= -80 - (-97 \text{ dBm}) \qquad (7.11)$$
$$= 17 \text{ dB}$$

Secondly, consider the receive level at Y from the transmitter at Z. The carrier level C_{YZ} is given as C_{YZ} (unfaded) = −45 dBm. Using (7.9) we can calculate C_{YZ} (faded)

$$C_{YZ} \text{ (faded)} = -45 \text{ dBm} - 40 \text{ dB} \qquad (7.12)$$
$$= -85 \text{ dBm}$$

The interference signal at Y from X (I_{YX}) can be calculated using (7.10)

$$I_{YX} = C_{YX} - F/B \text{ (Y antenna to Z)}$$
$$= -40 \text{ dBm} - 52 \text{ dB} \qquad (7.13)$$
$$= -92 \text{ dBm}$$

The carrier (C_{YZ}) to interference (I_{YX}) ratio is

$$C/I \text{ (faded)} = -85 - (-92 \text{ dBm}) \qquad (7.14)$$
$$= 7 \text{ dB (insufficient)}$$

The result of (7.14) shows that the interference is unacceptably high. To improve the situation let us first try to balance the receive levels at Y. First, increase the antenna at site Z to a 3-m dish (gain A_e increases by 5 dB). The unfaded carrier level is unchanged, hence C_{YX} (unfaded) = −40 dBm. Using (7.9) we can calculate the faded carrier, hence

$$C_{YX} \text{ (faded)} = -40 \text{ dBm} - 40 \text{ dB} \qquad (7.c)$$
$$= -80 \text{ dBm}$$

Now consider the interference signal at Y from Z. C_{YZ} now increases by 5 dB due to the 3-m dish, hence C_{YZ} = −40 dBm. Using (7.10) we can calculate the new interference level

$$I_{YZ} = C_{YZ} - F/B \text{ (Y antenna to X)}$$
$$= -40 \text{ dBm} - 52 \text{ dB} \qquad (7.15)$$
$$= -92 \text{ dBm}$$

The new carrier (C_{YX}) to interference (I_{YZ}) ratio is

$$C/I \text{ (faded)} = -80 - (-92 \text{ dBm}) \qquad (7.16)$$
$$= 12 \text{ dB (insufficient)}$$

The result of (7.16) shows that the C/I ratio is still insufficient since a minimum of 15 dB is required. To determine the interference from the other hop let us, once again, consider the receive level at Y from the transmitter at Z. The carrier level has increased to C_{YZ} (unfaded) of −40 dBm, as shown in (7.15). Using (7.9) we can derive the carrier level. Thus,

$$C_{YZ} \text{ (faded)} = -40 \text{ dBm} - 40 \text{ dB} \qquad (7.d)$$
$$= -80 \text{ dBm}$$

The interference signal at Y from X is as derived from (7.13). Hence, $I_{YX} = -92$ dBm. The carrier (C_{YZ}) to interference (I_{YX}) ratio is

$$C/I \text{ (faded)} = -80 - (-92 \text{ dBm}) \qquad (7.17)$$
$$= 12 \text{ dB (better, but still insufficient)}$$

This has balanced the C/I ratios (compare the results of (7.16) and (7.17)) with interference at Y from X much improved (7 dB to 12 dB), but it is still not good enough. Balancing the receive levels is often made out to be the panacea to all frequency-planning problems but the preceding example shows that it does not fix all problems. In the author's opinion its popularity stems mainly from the fact that it simplifies the calculations. On the negative side it can result in much larger antennas than would be required from a fading perspective, which can influence the project costs dramatically.

Let us now improve the situation by using the original configuration and just changing the Y antenna in the Z direction with a 3-m high-performance antenna with F/B of 70 dB with the same gain (45 dBi). The receive level at Y from the transmitter at X and the interference signal at Y from Z remain unchanged as derived in (7.11). The carrier (C_{YX}) to interference (I_{YZ}) is thus

$$C/I \text{ (faded)} = -80 - (-97 \text{ dBm}) \qquad (7.e)$$
$$= 17 \text{ dB}$$

The receive level at Y from the transmitter at Z remains unchanged using (7.12), hence

$$C_{YZ} \text{ (faded)} = -45 \text{ dBm} - 40 \text{ dB} \qquad (7.f)$$
$$= -85 \text{ dBm}$$

Now consider the interference signal at Y from X, with the high-performance dish using (7.13), hence

$$I_{YX} = C_{YX} - F/B \text{ (Y high performance antenna to Z)}$$
$$= -40 \text{ dBm} - 70 \text{ dB} \tag{7.g}$$
$$= -110 \text{ dBm}$$

The carrier (C_{YZ}) to interference (I_{YX}) ratio with the high-performance dish is thus

$$C/I \text{ (faded)} = -85 - (-110 \text{ dBm}) \tag{7.h}$$
$$= 25 \text{ dB}$$

This is a vast improvement and well within the 15-dB requirement, proving that high-performance antennas are the best approach to interference problems.

7.7.4.2 Overshoot Interference Example

Consider a typical radio route where overshoot is possible, as shown by Figure 7.17.

Tutorial Problem

Assume the transmit output power is +30 dBm, the receiver threshold is −80 dBm, and minimum carrier to interference ratio (C/I_{min}) of 15 dB. Also assume that all antennas are grid antennas. Typical RPEs for the 3-m and 4-m dishes are given in Figures 7.18 and 7.19. The objective of this tutorial is to determine the unfaded C/I ratio at site 4 at the input to the feeder cable behind the dish. Second, the effect of fading by 30 dB on hop 3-4 is analyzed. Finally, assume you wanted to double the capacity of hop 3-4 without changing equipment or frequency bands. You decide to replace the grids, with dual-polar feed solid dishes. Assume the same electrical parameters (for the sake of

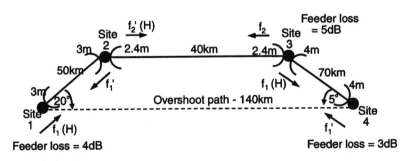

Figure 7.17 Overshoot interference.

Frequency Band (MHz) 1700 to 1900

Diameter: 3m

Gain: Single Polarized 32,5 ± 0.2 dBi at 1800 Mhz

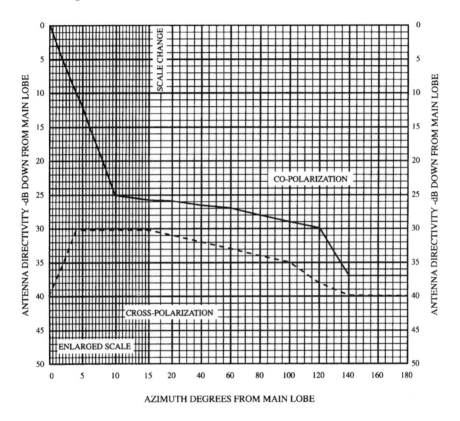

Figure 7.18 Typical RPE for 3-m grid dish.

the example) and determine the C/I ratio at site 4, ignoring any overshoot from the other sites.

Tutorial Worked Solution

The receive carrier level (unfaded) at site 4 (C_{RX4}) can be expressed as

$$C_{RX4} \text{ (unfaded)} = TX_3 - L_3 + A_{e3}(0) - FSL(70 \text{ km}) + A_{e4}(0)$$

$$(7.18)$$

where TX_3 is the transmit output power at site 3, L_3 is the feeder loss at site 3, $A_{e3}(0)$ is the antenna gain at site 3 on boresight, $A_{e4}(0)$ is the

Frequency Band (MHz) 1700 to 1900

Diameter: 4m

Gain: Single Polarized 35,4 ± 0.2 dBi at 1800 Mhz

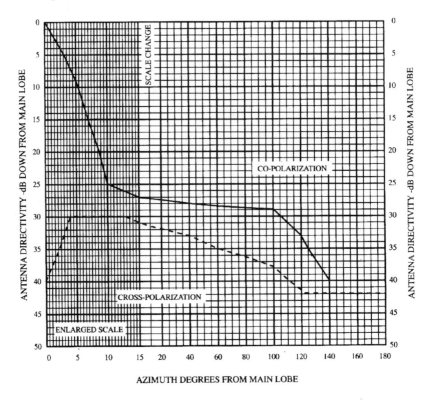

Figure 7.19 Typical RPE for 4-m grid dish.

antenna gain at site 4 on boresight, and FSL denotes the FSL (92.4 dB + $20 \log f$ GHz + $20 \log d_{km}$).

Using the tutorial values in (7.18) yields

$$C_{RX4} = +30 \text{ dBm} - 5 \text{ dB} + 35.4 \text{ dBi} - 134.4 + 35.4 \text{ dBi} \qquad (7.i)$$
$$= -38.6 \text{ dBm (unfaded)}$$

The interference level at site 4 (I_{RX4}) can be expressed as

$$I_{RX4} = TX_1 - L_1 + A_{e1}(20°)_{CO} - FSL(140 \text{ km}) + A_{e4}(5°)_{CO} \qquad (7.19)$$

where TX_1 is the transmit output power at site 1; L_1 is the feeder loss at site 1; $A_{e1}(20)$ is the antenna gain at site 1, 20 degrees off-boresight, copolar; $A_{e4}(0)$ is the antenna gain at site 4, 5 degrees off-boresight, copolar; and $FSL = 92.4$ dB $+ 20 \log f$GHz $+ 20 \log d_{km}$.

Using the tutorial values in (7.19) yields

$$I_{RX4} = +30 \text{ dBm} - 4 + (32.5 - 26) - 140.4 + (35.4 - 9) \qquad (7.j)$$
$$= -81.5 \text{ dBm}$$

The carrier (C_{RX4}) to interference (I_{RX4}) ratio (copolar) in an unfaded condition is thus

$$C/I_{CO} = -38.6 - (-81.5) \qquad (7.k)$$
$$= 42.9 \text{ dB}$$

If hop 3-4 fades by 30 dB, the received carrier level at 4 would be as

$$C_{RX4} = -38.6 \text{ dBm (unfaded)} - 30 \text{ dB} \qquad (7.20)$$
$$= -68.6 \text{ dBm (faded)}$$

In practice, for paths nearly parallel, there would be some correlation of fading; however, for a worst case scenario we assume the interference signal is unaffected by the fading, hence

$$I_{RX4} = -81.5 \text{ dBm} \qquad (7.l)$$

The copolar carrier (C_{RX4}) to interference (I_{RX4}) ratio with fading is

$$C/I_{CO} \text{ (faded)} = -68.6 - (-81.5) \qquad (7.m)$$
$$= 12.9 \text{ dB (insufficient)}$$

This is less than 15 dB, which is insufficient. To improve it, we need to change polarization. Re-assign hop 3-4 to vertical polarization. The wanted carrier level at 4 in a faded condition remains unchanged, as shown by (7.20), hence

$$C_{RX4} = -68.6 \text{ dBm (faded)} \qquad (7.n)$$

The interference level at site 4 (I_{RX4}) can be now be expressed as the sum of horizontal and vertical components

$$I_{RX4} = 10 \log ((I_{RX4}(H)/10) + 10 \log (I_{RX4}(V)/10)) \qquad (7.21)$$

Each one (H/V) can be expressed as

$$I_{RX4} (H/V) = TX_1 - L_1 + A_{e1}(20°)_{CO/XPOL} \qquad (7.22)$$
$$- FSL (140 \text{ km}) + A_{e4}(5°)_{XPOL/CO}$$

where TX_1 is the transmit output power at site 1, L_1 is the feeder loss at site 1,

$$A_{e1}(20°)_{CO/XPOL} = \text{Ant. gain (site 1), at } 20° - \text{Co/Cross polar} \qquad (7.oa)$$
$$= \text{Ant. gain } (0°) - \text{Antenna discrimination } (20°)$$

which is read off the antenna RPE curves,

$$A_{e4}(5°)_{XPOL/CO} = \text{Ant. gain (site 4), at } 5° - \text{Cross/Copolar} \qquad (7.ob)$$
$$= \text{Ant. gain } (0°) - \text{Antenna discrimination } (20°)$$

which is also read off the antenna RPE curves, and

$$FSL = 92.4 \text{ dB} + 20 \log f_{GHz} + 20 \log d_{km} \qquad (7.oc)$$

Using (7.22) we can derive the horizontal component of the interference at site 4

$$I_{RX4} (H) = TX_1 - L_1 + A_{e1}(20°)_{CO} - FSL (140 \text{ km}) + A_{e4}(5°)_{XPOL}$$
$$= +30 \text{ dBm} - 4 + (32.5 - 26) - 140.4 + (35.4 - 30) \qquad (7.23)$$
$$= -102.5 \text{ dBm}$$

Using (7.22) we can also derive the vertical component of the interference at site 4

$$I_{RX4} (V) = TX_1 - L_1 + A_{e1}(20°)_{XPOL} - FSL (140 \text{ km}) + A_{e4}(5°)_{CO}$$
$$= +30 \text{ dBm} - 4 + (32.5 - 31) - 140.4 + (35.4 - 9) \qquad (7.24)$$
$$= -86.39 \text{ dBm}$$

The total interference is the sum of the two I_{RX4} (H) and I_{RX4} (V)

$$I_{RX4} = 10 \times \log(10^{-86.5/10} + 10^{-102.5/10}) \qquad (7.25)$$

$$= -86.39 \text{ dBm}$$

As can be seen, the overall interference is determined by the vertical component (the horizontal being negligible in comparison). The overall C/I is the ratio of the carrier (C_{RX4}) and interference (I_{RX4}), hence

$$C/I = -68.6 - (-86.39) \text{ dB} \qquad (7.p)$$

$$= 17.8 \text{ dB}$$

It can be seen that by making the link cross-polar the interference is now acceptable because it is above the minimum requirement of 15 dB for this equipment. Note, however, that the maximum cross-polar protection occurs at the boresight of the antenna and deteriorates progressively until it virtually disappears at the back of the antenna. The final part of the example is to assess the effect of doubling the capacity by changing the grids with dual-polar solid dishes and using both polarizations.

The carrier level remains the same for both polarizations, hence

$$C_{RX4} \text{ (unfaded)} = -38.6 \text{ dBm} \qquad (7.q)$$

$$C_{RX4} \text{ (faded)} = -68.6 \text{ dBm}$$

The interference on the receiver C_{RX4} (H and V) from the oppositely polarized carrier is just attenuated by the cross-polar discrimination of the antenna

$$I = C \text{ (unfaded)} - XPD$$

$$= C \text{ (unfaded)} - 40 \qquad (7.26)$$

$$= -38.6 \text{ dBm} - 40 \text{ dB}$$

$$= -78.6 \text{ dBm}$$

The carrier (C_{RX4}) to interference (I) ratio is thus just the cross-polar discrimination

$$C/I \text{ (unfaded)} = 40 \text{ dB} \qquad (7.r)$$

$$= XPD$$

Including fading,

$$I = C \text{ (unfaded)} - \text{fade margin} - XPD$$
$$= C \text{ (faded)} - 40 \text{ dB} \qquad (7.27)$$
$$= -68.6 \text{ dBm} - -40 \text{ dB}$$

The C/I ratio in a faded condition (faded) is

$$C/I = C \text{ (faded)} - I$$
$$= C \text{ (faded)} - [C \text{ (faded)} - 40 \text{ dB}] \qquad (7.s)$$
$$= 40 \text{ dB}$$

The C/I is thus still 40 dB, which is the XPD figure. It can be seen that the C/I ratio in both cases is equal to the XPD of the antenna. Since the interference path is over the same hop as the carrier, the level of fading is the same. The only effect of the fading is to slightly rotate the carriers, thus degrading the XPD. Modern equipment's employ XPICs to reduce this effect.

7.7.5 Overshoot from Back-to-Back Antenna Systems

Back-to-back antenna systems are very easy to design from a link performance point of view, as they merely represent an additional insertion loss in the path power budget; however, they suffer from interference problems from the overshoot path if the overall diffracted path is not thoroughly blocked under all k conditions. The level of interference experienced is not intuitive and so it is easy to assume that no problems will exist only to find later that the interference is greater than expected. Just having the passive legs off-line or running the links on opposite polarizations will not necessarily solve the problem as illustrated in the worked examples in this section. The radio planner is encouraged to carefully consider the examples presented in this section before installing a back-to-back repeater.

7.7.5.1 Back-to-Back Interference

Consider a back-to-back link, with the two paths operating on opposite polarizations, with the site placed directly in-line in order to maximize the XPD of the antennas. The situation is represented diagrammatically in Figure 7.20.

Assume the XPD is 30 dB. The first (wanted) path is the path via the back-to-back antenna system. It experiences two FSL, less the back-to-back antenna gain. Typically the overall insertion loss is 10 dB to 20 dB more than the end-to-end FSL. The overshoot (unwanted) path experiences the end-to-end FSL plus the diffraction loss plus the XPD loss (30 dB). If the diffraction

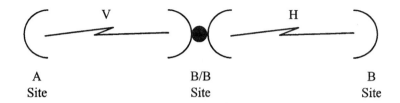

A
Site

B/B
Site

B
Site

Figure 7.20 Back-to-back interference.

loss is small, it means the C/I is only 10 dB to 20 dB, which in many cases is inadequate.

Trying to reduce overshoot interference by taking the passive site off the direct path introduces new problems. First, it results in the passive path lengths increasing. This increases the FSL, which are already very high. Second, although one gets sidelobe discrimination from the antennas, the XPD reduces dramatically.

On balance, despite the problems associated with placing the passive site off the direct path, this is the preferred solution. This will be demonstrated comprehensively in the following tutorial.

7.7.5.2 In-Line Passive

Assume that the passive site has been chosen in-line to minimize hop lengths and maximize XPD on the end-site antennas. Assume the end-to-end distance is 10 km and that the passive is centered in the middle of the path. Assume that there is no LOS between the end points. Further, assume the link is operating at 2 GHz, that the maximum size antennas are used throughout (4m) with gain of 35.4 dBi, and that the XPD at boresight is 30 dB. A side view of the site layouts is shown in Figure 7.21.

Consider the situation from Site A to Site C. The FSL can be calculated from (7.8). The overall FSL over path A–C using (7.8) is thus

$$FSL_{AC} \ (10 \ km) = 118.4 \ dB \tag{7.t}$$

The FSL over the two passive paths using (7.8) is

$$FSL_{AB} \ (5 \ km) = FSL_{BC} = 112.4 \ dB \tag{7.u}$$

Common parameters need not be considered in C/I calculations. The transmit output power is thus irrelevant. The wanted receive signal at *C*, ignoring common parameters, is

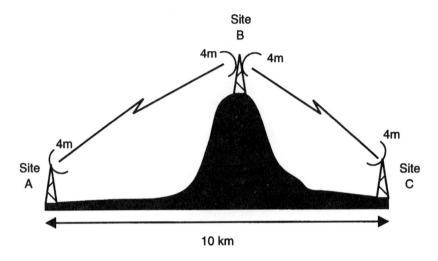

Figure 7.21 Side view of a back-to-back link.

$$C = A_e \times 4 - (FSL_{AB} + FSL_{BC})$$
$$= 35.4 \text{ dBi} \times 4 - (FSL_{AB} + FSL_{BC}) \qquad (7.28)$$
$$= -83.2$$

The interference level at C can be derived from

$$I_C = 10 \log((I_C(H)/10) + 10 \log(I_C(V)/10)) \qquad (7.29)$$

Each one (H/V) can be expressed as

$$I_C (V/H) = A_A(0°)_{CO/XPOL} - FSL_{AC} (10 \text{ km}) - DL + A_C(0°)_{XPOL/CO} \qquad (7.30)$$

where DL denotes diffraction loss. The vertical component of the interference at Site C can be derived from (7.30) as

$$I_C (V) = A_A(0°)_{CO} - FSL_{AC} (10 \text{ km}) - DL + A_C(0°)_{XPOL}$$
$$= (35.4) - 118.4 - DL + (35.4 - 30) \qquad (7.31)$$
$$= -77.6 - DL$$

The horizontal component of the interference at Site C can also be derived from (7.30)

$$I_C \text{ (H)} = (35.4 - 30) - 118.4 - DL + 35.4 \qquad (7.32)$$
$$= -77.6 - DL$$

The total interference is thus the sum of the horizontal and vertical components

$$I_C = 10 \times \log(10^{(-77.6-DL)/10} + 10^{(-77.6-DL)/10}) \qquad (7.33)$$
$$= -74.6 - DL$$

The carrier (*C*) to interference (*I*$_C$) ratio is thus

$$C/I = -83.2 - (-74.6 - DL) \qquad (7.v)$$
$$= -8.6 + DL$$

To achieve a C/I$_{\min}$ of 30 dB means that nearly 40 dB of diffraction loss is required. This is also required at a high value of *K* such as 5 and not at the traditional 2/3. To achieve this the path must be thoroughly blocked. It can also be seen that the 30 dB XPD did not translate into C/I.

7.7.5.3 Orthogonal Passive

Let us assume we now found a site at 90 degrees to the one end. This increases the one leg to 11.2 km. The site geometry is shown in Figure 7.22.

Consider the situation again from site A to site C. The FSL can be calculated from (7.8). The overall FSL over path A–C is

$$FSL_{AC} \text{ (10 km)} = 118.4 \text{ dB} \qquad (7.w)$$

The FSL over paths A–B and B–C, respectively, are

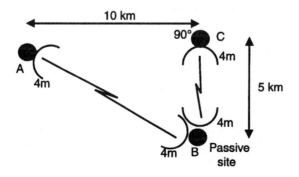

Figure 7.22 Passive site geometry.

$$FSL_{AB} \ (11.2 \ km) = 119.4 \ dB \qquad (7.x)$$
$$FSL_{BC} \ (5 \ km) = 112.4 \ dB$$

The wanted receive signal at C, ignoring common parameters, can be derived from (7.28). Hence

$$C = 35.4 \ dBi \times 4 - (FSL_{AB} + FSL_{BC})$$
$$= 141.6 - (119.4 + 112.4)$$
$$= -90.2$$

The interference level at C can be derived from (7.29). Each one (H or V) can be derived from (7.30). Ignoring common parameters we can derive

$$I_C \ (V/H) = A_A(27°)_{CO/XPOL} - FSL_{AC} \ (10 \ km) - DL + A_C(90°)_{XPOL/CO}$$
$$(7.34)$$

where DL is the diffraction loss. The vertical component of the interference at site C can also be derived from (7.31)

$$I_C \ (V) = A_A(27°)_{CO} - FSL_{AC} \ (10 \ km) - DL + A_C(90°)_{XPOL}$$
$$(7.35)$$

Figures 7.20 and 7.21 can be used to read off the CO and XPOL antenna discriminations. Thus,

$$I_C \ (V) = (35.4 - 27.5) - 118.4 - DL + (35.4 - 37) \qquad (7.z)$$
$$= -112.1 - DL$$

The horizontal component of the interference at site C can also be derived from (7.31)

$$I_C \ (H) = (35.4 - 37) - 118.4 - DL + 35.4 - 27.5 \qquad (7.36)$$
$$= -112.1 - DL$$

The total interference is the sum of the horizontal and vertical components

$$I_C = 10 \times \log(10^{(-112.1-DL)/10} + 10^{(-112.1-DL)/10}) \qquad (7.37)$$
$$= -109 - DL$$

The carrier (C) to interference (I_C) ratio is thus

$$C/I = -90.2 - (-109 - DL) \qquad (7.AA)$$
$$= 18.8 + DL$$

To achieve a C/I_{min} of 30 dB still means that more than 10 dB of diffraction loss is required at a high value of K such as 5. The situation has improved, but the path must still be thoroughly blocked.

7.8 Antenna Considerations

The choice of antenna is critical in the interference analysis of a microwave radio route. For frequency re-use it is the F/B ratio that is the critical parameter, and for overshoot protection it is the sidelobe and cross-polar discrimination aspects that are important. The F/B ratio is defined as the ratio of the gain in the desired forward direction to the gain in the opposite direction out of the back of the antenna. High-performance antennas have excellent F/B ratios, typically 10 dB to 20 dB better than standard antennas. High-performance antennas are significantly more expensive however, and add complexity and cost to the rigging. The towers may also need to be stronger to support the extra weight and windloading. The frequency re-use requirements thus have to be balanced with the other project requirements.

7.9 Intermodulation Products

An additional interference problem that is significant at VHF and UHF frequencies is caused by IMPs. Any two frequencies that are allowed to mix (beat) in a nonlinear device will generate an additional set of frequencies. These additional frequencies are called IMPs. If these products happen to fall within the pass band of the receiver, then an IMP problem exists.

The first source of IMPs is generated by the transmitter itself (including spurious and harmonics) and caused by the nonlinearity of the transmitter. They can be reduced by the transmit branching filters. The intermodulation products that are normally considered are those resulting from two carrier frequencies rather than those generated internally to the equipment. The ITU plans should ensure that the frequencies chosen minimize this effect, but different systems at a site can result in problem IMPs.

The second source of intermodulation products are generated within the receiver input stage due to the nonlinearity of the mixers, for example. Two off-channel signals can be mixed in the receiver to produce an IMP that corresponds to the wanted receiver frequency. This effect can be reduced by having narrowband receive filters.

Finally IMPs can be generated by any nonlinear physical device at the site, such as a rusty tower and fence. This is called the "rusty bolt effect." Metallic surfaces that have been oxidized behave in a nonlinear way like a diode. The metal structure can then become a radiating device (antenna), resulting in the IMPs.

If one mixes (modulates) one frequency (A) with another (B), upper and lower side bands $nA + mB$ and $nA - mB$ are produced. Examples of this are $A + B$, $A - B$, $2A + B$, $2A - B$.

The order of the products is determined by $n + m$. Hence $n + m = 2$ are second order and $n + m = 3$ are third order. Second-order (and further even-order) products are not problematic since the resulting frequency lies well off the receiver frequency. For example, if the two carrier frequencies 402 MHz and 407 MHz mix, the result is (402 + 407 = 809 MHz) and (407 − 402 = 5 MHz). Third-order (and further odd-order) products can cause problems since the new frequency may lie well within the receiver pass band. For example, $2 \times 402 - 407 = 397$ MHz, which is only 5 MHz away from 402 MHz.

Care must be taken to avoid third- and fifth-order products at a site when frequencies are allocated. Strict adherence to the ITU frequency plans should be maintained because they are designed for minimum IMP interference.

References

[1]   ITU-R F.385-6, Geneva, 1994.

8

Link Design

8.1 Introduction

In the previous chapters we studied the fading mechanisms due to propagation anomalies, interference effects, and equipment and antenna characteristics. In this chapter we will apply this knowledge to perform the design calculations for a radio link. The outages that are calculated need to be compared to the quality objectives that we studied earlier. In that chapter we learned that the availability requirements refer to events that occur for longer than 10 sec whereas the performance requirements refer to events that last less. Because ducting, diffraction fading, and rain fading are slow events (i.e., last longer than 10 sec—usually some hours or more) they need to be considered from the availability point of view. Multipath fading, however, is a fast event and needs to be considered from the performance point of view. We start by analyzing multipath fading outages, which can be compared to performance standards such as G.821 or G.826.

8.2 Multipath Fading Mechanism

Multipath fading is a complex fading mechanism, especially on wideband systems. Since it is difficult to visualize the problem some mathematics will be used to illustrate the problem.

8.2.1 Multipath Channel Transfer Function

The first step in analyzing the problem is to derive the transfer function of the atmosphere under multipath conditions. The transfer function is a mathematical concept used to analyze the effect of a signal through a network. It is represented by the output of the signal divided by the input.

In a multipath condition we input a microwave signal into the channel and the output consists of a direct signal and a delayed signal that has been reflected or refracted. Assume that the direct ray is represented by

$$a_1(t) = a_1 \exp(j\omega t) \tag{8.1}$$

where a_1 is the amplitude of the direct ray and $\omega = 2\pi f$ where f is the frequency.

Now let the delayed signal have a delay τ and phase ϕ, where τ represents the time delay of the echo, and ϕ represents the phase rotation of the signal. Then

$$a_2(t) = a_2 \exp(j(\omega(t - \tau) - \phi)) \tag{8.2}$$

where a_2 is the amplitude of the delayed (secondary) ray.

The sum of these two at the receive antenna is represented by $x(t)$

$$x(t) = a_1(t) + a_2(t) \tag{8.3}$$

The transfer function $H(\omega)$ is represented by the output over the input. Hence, dividing (8.3) by (8.1) yields

$$H(\omega) = x(t)/a_1(t) \tag{8.4}$$

$$= 1 + (a_2/a_1) \exp(-j(\omega\tau + \phi)) \tag{8.5}$$

If we call the relative amplitude difference between the echo and the direct ray "b" (8.5) can be written as

$$H(\omega) = 1 + b \exp(-j(\omega\tau + \phi)) \tag{8.6}$$

where b is the amplitude of the echo relative to the direct ray and is always greater than zero except for the case where the echo has an amplitude greater than the direct ray and its value is greater than unity. τ is the value of the delay of the echo. Although it is normally positive, if the echo arrives before the main signal its value is negative.

Equation (8.6) describes the channel transfer function using two-ray interference analysis, however, it is also known that the overall signal level is also attenuated in a real channel. At times the fading results in a flat attenuation across the frequency band of interest, and this can be represented by an attenuation factor α that is independent of frequency. Hence (8.6) can be written as

$$H(\omega) = \alpha(1 + b \exp(-j(\omega\tau + \phi))) \tag{8.7}$$

We can now convert (8.7) from its complex representation into a Cartesian representation. To explain the concepts of complex and Cartesian equations, the generic conversion is

$$Z = |Z|e^{j\varphi} = a + jb \tag{8.8}$$

where $a = |z|\cos\varphi$ and $b = |z|\sin\varphi$, where $|z| = \sqrt{(a^2 + b^2)}$ and denotes the magnitude of the signal and $\varphi = \arctan(b/a)$ and is the phase of the signal. The Cartesian representation of (8.7) is

$$H(\omega) = \alpha[(1 + b\cos(\omega\tau + \phi)) - j(b\sin(\omega\tau + \phi))] \tag{8.9}$$

To determine the amplitude versus frequency response of the transfer channel we need to determine the magnitude of (8.9). This is plotted in Figure 8.1 and is derived from

$$|H(\omega)| = \alpha\sqrt{(1 + b^2 + 2b\cos(\omega\tau + \phi))} \tag{8.10}$$

8.2.2 Frequency Response

Equation (8.10) is a periodic function that has minima when $\cos(\omega\tau + \phi) = -1$. Hence,

$$(\omega\tau + \phi) = \pi \pm 2\pi n \quad \text{for} \quad n = 0, 1, 2 \ldots \tag{8.a}$$

The difference between two notches can be derived as follows:

$$\omega_n\tau = \begin{cases} \pi - \phi & \text{for } n = 0 \\ 3\pi - \phi & \text{for } n = 1 \end{cases} \tag{8.b}$$

The difference between two notches is

$$\omega_1\tau - \omega_0\tau = 2\pi \tag{8.11}$$

Considering that $\omega = 2\pi f$, we can rearrange (8.11) to derive the very important result

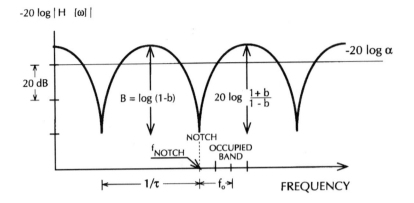

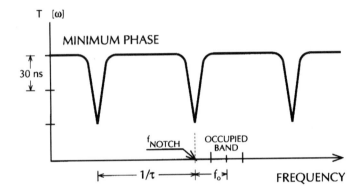

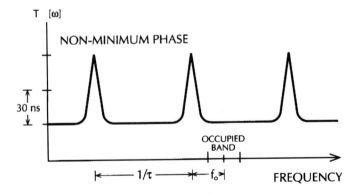

Figure 8.1 Frequency response and group delay of multipath channel.

$$\Delta f = 1/\tau \qquad (8.12)$$

In other words, the difference in frequency between two minima (notches) is equal to the inverse of the delay of the echo caused by that notch, as shown by Figure 8.1. If the delay varies with time over a multipath channel as is the case with a microwave link during ducting conditions, the notch will move across the bandwidth of the receiver.

8.2.3 Minimum and Nonminimum Phase Conditions

The multipath channel is a function of both the delay of the secondary signal (τ) and its relative amplitude value (b). The combination of these two values results in different conditions.

When the relative amplitude of the echo is less than the main beam and the delayed signal occurs after the main signal or when the relative amplitude of the echo is greater than the main beam and the delayed signal occurs before the main beam, the condition is known as a minimum phase condition. Under opposite circumstances it is known as a nonminimum phase condition. This is summarized in Table 8.1.

The reference to minimum phase and nonminimum phase relates to circuit theory where the transfer function is drawn (via the Laplace transform) in the s-plane. Maxima and minima in this case are represented by poles and zeros, respectively. The transfer function of the multipath channel can be drawn on the s-plane as an infinite series of zeros (corresponding to the notch minima), as shown in Figure 8.2.

It can be seen from Figure 8.2 that for any given frequency ω, the phase τ is always less in the minimum phase condition (left-hand side), hence its name. As one increases the frequency, the phase increases for the minimum phase condition but decreases for the nonminimum phase condition, which means that the amplitude response for the two cases is identical but the group delay response is inverted (i.e., one is a mirror image of the other).

The maximum value possible is $\alpha(1 + b)$ and the minimum value possible is $\alpha(1 - b)$. In other words the closer b is to unity the deeper is the notch. The ratio of the minimum to maximum values expressed in decibels is

Table 8.1
Minimum and Nonminimum Phase Conditions

Minimum phase	$\tau > 0$	$0 < b < 1$
	$\tau < 0$	$b > 1$
Nonminimum phase	$\tau > 0$	$b > 1$
	$\tau > 0$	$0 < b < 1$

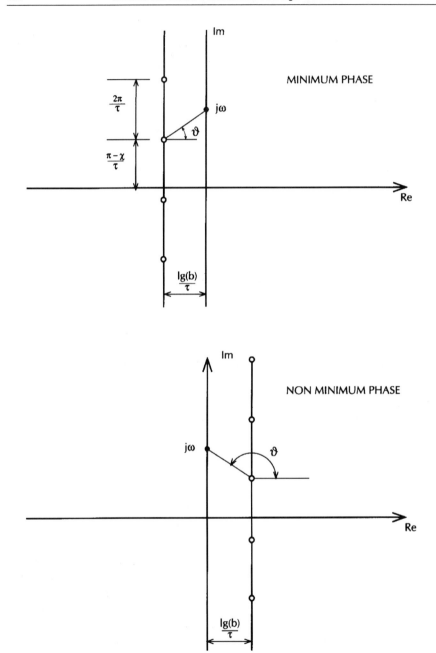

Figure 8.2 Pole zero diagram of phase condition.

$$H_{dB} = 20 \log((1 - b)/(1 + b)) \qquad (8.13)$$

Also, since minima are separated by $1/\tau$, the longer the delay the closer together the notches are in frequency.

8.2.4 Group Delay

The phase response can be derived from (8.9) as

$$\varphi = \arctan((b \cdot \sin(\omega\tau + \phi))/(1 + b \cdot \cos(\omega\tau + \phi))) \qquad (8.14)$$

The group delay is defined as the rate of change of phase with frequency, hence it can be derived from the derivative of (8.14) with respect to frequency

$$T(\omega) = d\varphi/d\omega \qquad (8.15)$$

If one performs this derivative and plots the response one obtains the curve for the minimum and nonminimum phases, respectively, as shown in Figure 8.1. The amplitude response is identical in both cases.

8.2.5 Mathematical Models

As previously mentioned the transfer function of the multipath channel is dependent on four variables: α is the attenuation factor constant across the receiver bandwidth, b is the relative amplitude of the delayed signal to the main one, τ is the propagation delay of the echo, and ϕ is the phase shift of the echo.

As far as these parameters are concerned, they are only known in statistical terms from field measurements. Because one frequency response can be reconstructed in many different combinations of these four parameters, a mathematical description that is based on physical reality is almost impossible. The only method that has been used to analyze the problem is to produce a mathematics model that assumes one of the variables constant.

In reality, as many as 11 rays can be present in an atmospheric multipath situation [1]. The number of rays will always be an odd number, as shown by Figure 8.3.

For practical applications the number of rays may be reduced to three or less. Specifically, all the mathematical models are usually based on the three-ray model or a simplification thereof. In this model we have a direct ray and two reflected/refracted rays. One of these delayed rays has a delay time that is sufficiently small to assume that its contribution is coherent and, hence,

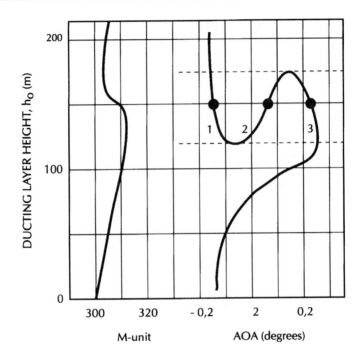

Figure 8.3 Effect of a duct on wavefront.

represents flat fading rather than frequency selective fading. Now, as we have already said, we need to reduce the number of parameters from four down to three for analysis purposes. To do this we need to assume one of the four parameters is constant. One such model assumes that the delay is constant and is based on work done by William Rummler of Bell Laboratories [2]. Using the propagation data from a 26.4-mi hop near Atlanta, Georgia (USA), the model was confirmed in 1977. It was found that for a channel with a 30-MHz bandwidth, a model with 6.3 ns of delay produced accurate results for all channel conditions.

Another model assumes the attenuation factor is constant. This is probably the most physically representative model because radio receivers are able via their AGC to compensate for flat fading. This is referred to as a two-ray model because the effect of the third ray producing the flat attenuation is eliminated.

8.3 Multipath Fading Outages

The outages due to multipath fading depend on parameters such as frequency, hop length, terrain type and roughness, climatic conditions, and path clearance.

The various semiempirical models developed to predict outages all use these parameters to some extent. Each country has unique parameters; therefore, no worldwide model has been found that accurately models the effects for all hops. Researchers are continually adjusting the formulae to more accurately match the hops that are being monitored. Some of these results can be traced by following the changes in formulas given in the various versions of ITU 530. Rounding off the ordinates used in the formulas, most of the results show that the outages are roughly proportional to frequency, distance cubed, and a geoclimatic factor determined from the hop terrain and climatic zone.

8.3.1 Flat Fading Outage

In the microwave context, flat fading is caused by multipath conditions where signal distortion effects can be ignored. This is typically the case for systems below about 8 Mbit/s, provided the hop lengths are short (less than 50 km). The effect of flat attenuation can easily be seen by plotting attenuation verses BER. It can be seen in a digital system that as signal attenuation is increased, the BER is initially unaffected until close to the receiver threshold where the BER increases steeply, as shown by Figure 8.4. This is called the threshold effect.

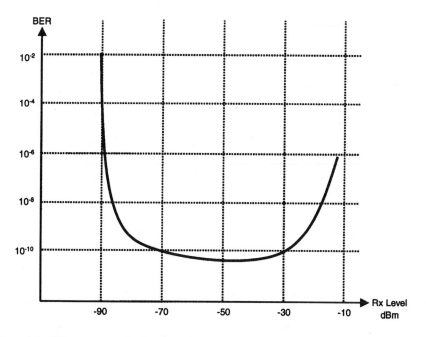

Figure 8.4 BER versus receive level.

This corresponds to the critical thermal noise level at the receiver input. Multipath fading causes a fast, deep signal attenuation that can cause an outage if the fade margin is exceeded. When designing a radio link it is important to know the probability of this event occurring relative to the depth of fading. Each link has a different inherent probability to be adversely affected by atmospheric fading depending on the length of the link, the frequency at which it operates, the path profile, and the climatic conditions. Years of testing on radio links (analog and digital) have shown that the probability distribution curve follows a Raleigh distribution for deep fades, which means that for every 10 dB of fading, the probability of occurrence decreases tenfold. Thus, there are 10 times more 10-dB fades than 20-dB fades. This fact means that the nature of fading is known for any hop—it is just the absolute level of fading that needs to be determined. One can express the Raleigh fading formula as

$$p(F < M) = P_0 \cdot 10^{-M/10} \tag{8.16}$$

In other words, the probability of a fade exceeding a set fade margin M is proportional to a set multipath fading occurrence factor P_0. This curve can be physically traced by monitoring the AGC level of a radio receiver. A fade analyzer records the number of times a signal fades to a specific level. The Raleigh curve can then be determined by plotting the receiver fade depth in decibels versus the probability that the ordinate is exceeded. The Raleigh curve is shown in Figure 8.5, where it will be seen that the first 15 dB or so of shallow fading displays a Gaussian distribution with the linear Raleigh portion occurring for deep fades.

Very deep fades will also deviate from the theoretical Raleigh curve because the number of values at this depth is often insufficient to be statistically reliable. By extrapolating the Raleigh distribution curve to intersect with zero-fade depth, the value of P_0 can be obtained. Since it is often not practical to measure this value, a theoretical value is required. A number of semiempirical formulae exist, such as the Vigants–Barnett formula and the formulas presented in ITU-R F.530.

8.3.2 Outage Predictions: Barnett-Vigants Model

A generic fading formula expresses the fading outage [3]

$$P(W) = K \cdot Q \cdot W/Wo \cdot f^B \cdot d^C \% \tag{8.17}$$

where $P(W)$ represents the percentage of time that the received power W is not exceeded, f is the frequency in gigahertz, d is the path length in kilometers,

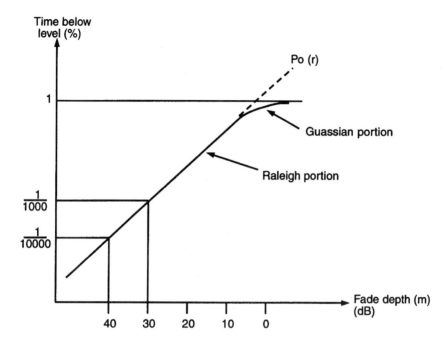

Figure 8.5 Raleigh distribution.

K is the geoclimate and terrain effects on climate, Q denotes the factor for variables other than those dependent on d and f, and Wo is the received power in nonfading conditions.

Using the Barnett-Vigants ordinates, this becomes

$$P(W) = K \cdot Q \cdot W/Wo \cdot f \cdot d^3 \qquad (8.18)$$

where

$$K \cdot Q = A \cdot 10^{-3}/S_1^{1.3} \quad \text{for } A = \begin{cases} 4.1 & \text{for coastal areas} \\ 3.1 & \text{for subtropical} \\ 2.1 & \text{for average terrain (flat)} \\ 1 & \text{for mountainous} \end{cases} \qquad (8.c)$$

and S_1 is the roughness of the terrain as measured by the standard deviation of the terrain elevations measured at 1-km intervals and excluding the radio site heights. $(6m < S_1 < 42m)$.

For narrowband systems we can write

$$(W/W_0) = 10^{-M/10} \qquad (8.19)$$

where M denotes the flat fade margin in decibels. Hence, the Barnett-Vigant method for determining the probability of exceeding a certain fade margin M is

$$p(F < M) = P_0 \cdot 10^{-M/10} \qquad (8.20)$$

where $P_0 = A \cdot f \cdot d^3 \, 10^{-3}/S_1^{1.3}$.

8.3.3 Outage Predictions: ITU 530-7 Method

Various formulas have been developed and presented by the ITU [4]. The formulas and methods presented are an attempt to define prediction models that allow one to accurately predict the outage time for any given hop. Because the fading is caused by multipath, one tries to predict the probability of multiple paths existing. Multipath is caused by ground reflections and atmospheric refractions under ducting or near-ducting conditions. The probability of the refractivity gradient exceeding -100 N-units/km (P_L) is thus relevant. The angle that a nonspecular ground reflection would have off the average profile, or the grazing angle, is also relevant. It is also well known that if the path cuts across the duct rather than being launched parallel to the duct, the effect of the duct is reduced. The path inclination is thus relevant. The geoclimatic factors are also relevant, because they will have an affect on the amount of inherent multipath fading a path could have. In the latest method [4] the geoclimatic factor K is calculated as a function of site location (site latitude, longitude) and ducting probability (P_L). Ignoring site latitude or longitude, this can be expressed as

$$K = 5 \times 10^{-7} \times 10^{-0.1(C_0)} P_L^{1.5} \qquad (8.21)$$

where C_0 is the terrain altitude coefficient and takes the values

$$C_0 = \begin{cases} 1.7 & \text{from 0–400m AMSL} \\ 4.2 & \text{from 400m–700m} \\ 8 & \text{above 700m} \end{cases} \qquad (8.d)$$

The path inclination can be calculated from

$$\epsilon_p = |hr - he|/d \qquad (8.22)$$

where *hr* and *he* are the heights of the transmitting and receiving antennas above sea level and *d* is the hop length in kilometers.

The average worst month fade probability, *Pw*, can thus be expressed as

$$Pw = K \cdot d^{3.6} \cdot f^{0.89}(1 + \epsilon_p)^{-1.4} \cdot 10^{-A/10} \% \qquad (8.23)$$

where K denotes the geoclimatic factor (worst month), d is the path length in kilometers, f is the frequency in gigahertz, ϵ_p is the path inclination in millirad (max value 24), and A denotes the fade margin in decibels. The dependence on the path profile expressed as roughness factor (S_1) or grazing angle ϕ in previous versions has been dropped in this method.

8.3.4 Selective Fading Outage

In relatively short (less than 30 km) low-capacity links (2, 4, and 8 Mb/s), the bandwidth of the receiver, relative to the inverse of the delay times of the additional rays caused by multipath conditions, is sufficiently small that the signal distortion (caused by the frequency selective nature of the fading) can be ignored. In medium- and high-capacity links (e.g., 34 or 155 Mb/s), this is not so. In fact, the primary source of outage in a high-capacity system is due to signal distortions, which result in severe intersymbol interference (ISI). Because the demodulator of a radio receiver is designed according to the Nyquist principle in that the decision point occurs when the tails of a previous pulse are at a minimum, ISI results in closure of the eye diagram and an outage condition (BER $> 10^{-3}$). Increasing the receiver signal level, for example by increasing the fade margin, does not help the situation because the eye closure is not caused by thermal noise but by distortions of the amplitude and group delay across the receiver bandwidth.

8.3.4.1 System Signature

There are various methods for predicting outage times such as using a linear amplitude distortion (LAD) approach, a time domain multiecho approach, normalized signatures, and composite fade margin methods. The normalized signature and the composite fade margin approaches will be discussed further. These methods provide outage predictions based on knowledge of how the radio equipment responds to the inband distortion caused by the multiple-ray interference. It was shown earlier that if two-ray interference is plotted with frequency, a notch occurs. This notch was shown for a specific delay value of the secondary ray. In a dynamic fading channel, the instantaneous changes in refractive gradients would mean that each ray would travel a slightly different path to the next ray if plotted in the time domain. If one imagines that the delay of the secondary ray is incrementally increased, the notch would thus move across the channel. In practice this is precisely what happens. The fading effect is due to a frequency notch that sweeps across the channel. The outage is entirely dependent on how the radio equipment responds to this dynamic distortion event.

The measurement of the equipment's response is called a signature curve. Both the normalized signature method and the composite fade margin methods rely on having an accurate demodulator signature curve for the radio equipment. The signature measurement is done by setting up a test bench, which simulates the delayed echo signal. An attenuator is introduced to simulate flat fading that would affect the direct path and the echo. Then the signal is split into a main path and a delayed path. The delayed path undergoes a constant delay using a fixed length of waveguide and the phase and amplitude of the echo is controlled in such a way that the notch can be visually seen on a spectrum analyzer. At the input, a pseudorandom bit pattern is sent and a BER tester is employed at the receiver to measure the error rate. The depth of the notch can be seen on the spectrum analyzer.

The frequency shift of the notch is achieved by varying the phase of the echo. More specifically, the frequency shift can be written as

$$\Delta f_{\text{NOTCH}} = \Delta\phi/2\pi\tau \tag{8.24}$$

For a fixed delay (τ = 6.3 ns), this shift should be approximately 0.44 1 MHz per degree. The amplitude of the echo, which would cause an outage, can be checked by varying the echo amplitude (α) until a certain BER condition, such as BER = 10^{-3}, is reached. This amplitude α can be converted to the frequency notch attenuation, B, by

$$B = -20 \log_{10}(1 - b) \text{ dB} \tag{8.25}$$

where b denotes the amplitude value of the delayed signal assuming the main signal has a value of unity.

The signature is thus measured by setting the phase in such a way that the notch has a certain distance from the band center, represented by f_0, and then varying the amplitude required to cause an outage (e.g., 10^{-3}) condition. (Note that the BER = 10^{-6} condition can also be used.) This is then repeated for various notch conditions across the bandwidth of the receiver such that the signature curve is produced s shown in Figure 8.6.

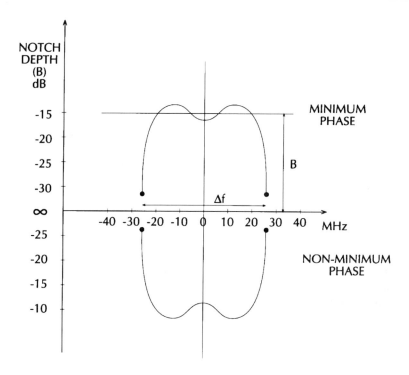

Figure 8.6 Typical signature curve.

8.3.4.2 Normalized Signature Outage Method

The normalized system signature method assumes that the outage is equal to the area under the average of the nonminimum phase and minimum phase signature curves.

This can be approximated by a rectangular area given as

$$K_n = T^2 \cdot W \cdot \lambda a / \tau_r \tag{8.26}$$

where T is the system baud period (nanoseconds), that is, symbol time; W is the signature width (gigahertz), λa is the average of signature depth $(1 - bc)$, and τ_r is the reference delay for λa (nanoseconds).

8.3.4.3 Composite Fade Margin Method

It was mentioned earlier that dispersive fading is not related to the received power level (thermal noise) but is the additional fading that caused errors due to distortion. If one knows the exact flat fading outage and has determined the dispersive nature of the path, then the dispersiveness of equipment characteristics can be determined from measurements. The dispersive fade margin is given as [5]

$$DFM = DFMR - 10 \log(DR/DR_0) \text{ dB} \qquad (8.27)$$

where $DFMR$ is the reference DFM, DR_0 is the reference dispersive ratio, and DR is the known path dispersive ratio.

Using the Rummler model, Bell Labs in the United States have developed a formula to derive the DFM from the signature curve

$$DFM = 17.6 - [10 \ln(Sw/158.4)] \text{ dB} \qquad (8.28)$$

where $Sw = e^{-B(\text{dB})/3.8} \, \Delta f \text{ MHz}$.

8.4 Rain Fading Outage

The propagation aspects of rain fading were discussed in the previous section. To compute the rain outage one needs to determine the rain rate required to attenuate the path such that the fade margin is exceeded and then determine how often this rain rate occurs for the geographical area under consideration. The ITU has produced a world map where rain regions have been defined according to rain rate. Region A has the lowest rainfall rate and Region P has the highest. The path attenuation that corresponds to a rainfall rate that is only exceeded for 0.01% of the time is expressed as

$$A_{0.01\%} = \gamma_R \cdot d \cdot r \text{ dB} \qquad (8.29)$$

where γ_R is the specific attenuation in dB/km, d is the hop length in kilometers, and r is the distance factor.

Very high rainfall rates tend to cover a smaller geographical area, hence the rainfall will not affect the entire hop length equally. The effective hop length is calculated as the product of the real hop length d and a distance factor r

$$r = 1/(1 + d/d_0) \qquad (8.30)$$

where $d_0 = 35e^{-0.015R(0.01\%)}$ for rain rates below 100 mm/hr.

The specific attenuation can be derived from the rain rate R (mm/hr) using

$$\gamma_R = kR^\alpha \tag{8.31}$$

The regression constants k and α can be obtained from ITU references [6]. For example, at 15 GHz, ignoring path inclination and polarization tilt, the factors for vertical polarization are $k = 0.0335$ and $\alpha = 1.128$.

The design method would be to increase antenna size on a hop until the fade margin equaled or exceeded the path attenuation A calculated previously. For example, for an availability outage objective of 99.99% ($u = 0.01\%$) in rain region E ($R0.01\% = 42$ mm/hr) on a 10-km 15-GHz link, the path attenuation can be calculated using (8.29) and (8.31), hence

$$A\,(0.01\%) = \gamma_R \cdot d \cdot r \text{ dB} \tag{8.e}$$
$$= kR^\alpha d \cdot r$$

Substituting actual values for R^α yields

$$R^\alpha = 42^{1.128} = 67.8 \tag{8.f}$$

and

$$r = 1/(1 + d/d_0) = 1/(1 + d/35e^{-0.015\,R(0.01\%)})$$
$$= 1/(1 + 10/35 \cdot e^{-0.015 \cdot 42}) \tag{8.g}$$
$$= 0.65$$

Hence,

$$A\,(0.01\%) = kR^\alpha d \cdot r$$
$$= 0.0335 \times 67.8 \times 10 \times 0.65 \text{ dB} \tag{8.h}$$
$$= 15 \text{ dB}$$

A fade margin of 15 dB is thus required to meet the outage objective of 99.99%. This fade margin is perfectly adequate for this link; however, in practice one may consider using a higher frequency band such as 23 GHz for this link, in which case a higher fade margin would be required.

For other unavailability percentages (p) between 0.001% and 1% the path attenuation is

$$Ap = A(0.01\%) \times 0.12p^{-(0.546+0.043\log p)} \tag{8.32}$$

8.5 Total Outage

In order to determine the overall performance outage one must combine the probability of outage due to flat and selective fading. There are several ways one can do this, but the most common is to use the composite fade margin and then to use this value with the formulas developed for narrowband fading that are included in ITU recommendation 530-7. Hence, the probability of having an outage (or more generally, of exceeding a certain BER threshold) can be represented by

$$P(BER > 10^{-n}) = P_0 \cdot 10^{-M/10} \qquad (8.33)$$

where M represents the composite fade margin. M is the composite fade margin, which is the sum of the thermal flat fade margin (FFM) and the signature-curve derived dispersive fade margin (DFM), expressed in decibels and normally assumes no interference. If interference is present one should worsen the threshold accordingly, as discussed in Chapter 7.

The availability outage is still treated separately by the ITU, however, it is not uncommon to convert the performance outage into a two-way annual figure and add it to the availability outage. A method of converting from average worst month figures to average annual figures is given in [4]. A logarithmic geoclimatic conversion factor ΔG is used to compute p (annual outage) from Pw (worst month outage)

$$p = 10^{-\Delta G/10} Pw \qquad (8.34)$$

Although it is thus possible to compute a combined outage, there are no ITU outage objectives against which to compare the result. In the United States the Bellcore standards define a reliability objective that is the combination of performance and availability objectives [7].

8.6 Countermeasures

Countermeasures against the effects of fading can be seen in three categories: system techniques, nondiversity techniques, and diversity techniques.

8.6.1 System Techniques

For flat fading it is sufficient to increase the available fade margin. This can take the form of increasing the system gain by using larger antennas, increasing the transmit output power, or increasing the receiver threshold level. All of these have their limits: Antennas are manufactured to a certain maximum size; transmit output power is limited due to distortion problems; and the receiver threshold is limited by the residual background thermal noise, which is proportional to bandwidth. (The wider the bandwidth the greater the thermal noise and the less the receiver threshold.) Once these factors have reached their limits, diversity is the only option to increase the performance. For selective fading the problem is not the receive field strength but distortion, and so equalization is the main system countermeasure that can be employed. Because this complex problem is mainly an equipment designer's problem rather than a system engineering problem, it will not be discussed in detail but the principles briefly discussed. Equalization can be effected at baseband or IF levels. The principle of the IF equalizer is to linearize the frequency response by adaptively creating a transfer function complementary to the actual transfer function of the channel. This can only directly affect the amplitude response, and therefore its performance against minimum phase and nonminimum phase fading is substantially different. It is not therefore capable of being a good enough countermeasure to selective fading. The physical circuitry normally takes the form of a slope or bump equalizer. Filters are used at different frequencies to sample the amplitude levels and, by means of feedback circuitry, the response is linearized. In other words, a frequency response with a positive slope is counteracted with a negative slope equalizer and a frequency response with a notch is compensated for with a complementary bump.

The most efficient system countermeasure against selective fading is the time domain-based baseband equalizer. In the undistorted state the tails of the pulse waveform ($\sin(x)/x$), as seen by the detector circuitry (demodulator eye-diagram), correspond with a minimum, in line with the Nyquist criteria. When distortion is present this component is no longer zero and ISI occurs. The objective is to subtract the interfering component from the signal. This is achieved by monitoring the postcursors (those tails from the preceding pulse) and precursors (the tails from the successive pulse) and applying the necessary delay times to ensure that the tails are zeroed at the point of reading. Figure 8.7 shows the improvement on the signature using various equalizers.

8.6.2 Nondiversity Techniques

Due to the financial implications of diversity, nondiversity techniques should first be considered: One such technique relies on the fact that only rays launched within approximately 0.5 degrees from the horizontal are subject to ducting. By placing the antenna much higher at one end than the other, a high–low arrangement can be obtained where fading is greatly reduced [8]. Another

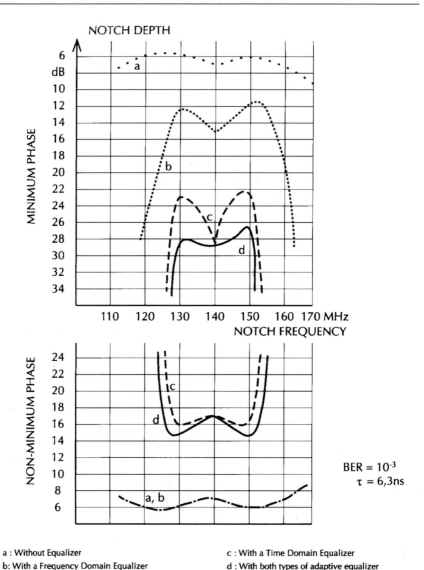

Figure 8.7 Equalizer signature improvement.

technique is to tilt the antenna slightly upward, causing a slight loss of signal under normal conditions but reducing the ground reflected ray (and hence the outage due to multipath) under fading conditions [9].

Some administrations position the antenna low, allowing a diffraction loss under normal conditions but at the same time ensuring that the reflected ray is blocked. This is a risky method as discussed elsewhere in this document because diffracted links may suffer severe outages due to low K values. Finally, an antireflecting antenna can be positioned such that reflections are canceled. The disadvantage of this method is that it becomes expensive. Antireflecting antenna systems also tend to have a high antenna loading due to the heavy tuning mechanisms required in order to optimize the spacing. This technique does, however, offer considerable advantages for a specular reflection problem.

8.6.3 Diversity Techniques

8.6.3.1 Comparison of Methods

Various forms of diversity are available to the path designer. The ITU [10] provides some useful data on different diversity schemes. Diversity schemes that can be used on point-to-point microwave links include

- Angle diversity;
- Space diversity with RF or IF combiners, which can be minimum dispersion or maximum power;
- Space diversity with baseband switching;
- Frequency diversity (inband or crossband; 1 + 1, or n + 1);
- Hybrid diversity (space diversity and frequency diversity with two or four receivers).

Angle diversity has been quoted in some literature as performing well against selective fading, however, it is not widely implemented yet because it has not been conclusively proved to be effective in practice.

Crossband frequency diversity is a very efficient method from a propagation point of view but is not very spectrum efficient because it requires that two frequency bands be available. One could use a high-frequency band such as 18 GHz as the protection channel assuming that when rain affects this band there is no multipath fading on the lower frequency band, say 6 GHz (the turbulent conditions associated with rain do not favor multipath). This would allow one to use the high-frequency band over much larger distances than usual. Although this may sound interesting in theory, it is unlikely to be used in practice because it is so wasteful of valuable frequency spectrum.

Inband frequency diversity is the most common form of diversity because when an n + 1 system is configured, one of the channels can be used for protection. A dedicated protection channel such as a 1 + 1 system is not as frequency efficient but affords a high level of protection. One can also put lower priority traffic on the protection channel that can be dropped when switching takes place, thus improving the spectral efficiency. Frequency diversity is not allowed in many countries due to the extra spectrum usage.

Space diversity is very spectrum efficient and provides excellent performance against multipath fading but is expensive. The concept is to separate the two antennas in the vertical plane such that when there is phase cancellation on the main path due to multipath fading, the diversity path is not affected due to the extra path length. Typically, provided there are at least two hundred wavelengths of separation between the antennas the two paths will not be correlated. The details of the space diversity equipment and branching arrangement were shown in Chapter 4, where it can be seen that two antennas and feeders are required at each end. The additional equipment cost is made worse by the fact that often stronger and higher towers are required. Another disadvantage of space diversity is that it does not offer a separate stand-by channel for maintenance purposes as frequency diversity does. The degree of improvement when using any of the diversity options depends on the amount of uncorrelation between the main channel and the diversity channel.

8.6.3.2 Frequency Diversity Outage

For frequency diversity the improvement factor is directly proportional to the frequency separation. The improvement factor [4] is given by

$$I_{FD} = 0.8/fd \cdot (\Delta f/f) \cdot 10^{F/10} \qquad (8.35)$$

where Δf is the frequency separation in gigahertz, f is the carrier frequency, d denotes the hop distance in kilometers, and F is the fade margin in decibels.

The outage P with frequency diversity is given by

$$P_{FD} = P/I_{FD} \qquad (8.36)$$

8.6.3.3 Space Diversity Outage

For space diversity using baseband switching the improvement factor is [11]

$$Isd = 1.2 \times 10^{-3}(f/d) \cdot s^2 \cdot \gamma^2 \cdot 10^{A/10} \qquad (8.37)$$

where s denotes the antenna separation in meters, γ_{dB} is the difference between the main and diversity receive levels in decibels ($20 \log(\gamma)$), f is the frequency in gigahertz, and d is the path length in kilometers.

The outage P with space diversity is given by

$$P_{SD} = P/I_{SD} \qquad (8.38)$$

The ITU [4] provides a method for calculating an improvement factor for systems using a phase combiner. Prediction methods for two- and four-receiver systems are also given.

Space diversity generally provides better improvement, in practice. If one equates the correlation factors for comparison purposes one can determine that at, for example 2 GHz, 10m of space diversity spacing is equivalent to 14 MHz of frequency separation. At 7 GHz, the same spacing is equivalent to 610 MHz of spacing. One can see therefore that inband frequency diversity (whose frequency spacing is limited by the ITU-R channel plans) is more efficient at lower frequencies. Tower height, on the other hand, can be a limiting factor for space diversity; and in the end a solution needs to be found depending on the particular situation rather than by rules of thumb. Despite this, it can generally be stated that at higher frequencies space diversity is more efficient than inband frequency diversity provided the spacing is not limited as shown previously. As a rule of thumb, the spacing of the antennas should be separated by 200 wavelengths to ensure the two signals are not correlated.

8.6.3.4 Hybrid Diversity

A cost-effective and very efficient method for 1 + 1 systems is hybrid diversity, where the frequency diversity switch is used to switch two channels separated spatially over the link. To achieve this, at one end two antennas are employed, each connected to the respective main and standby transmitters and receivers. At the far end one antenna is used but the receivers are switched by the frequency diversity switch. Space and frequency diversity are thus achieved in both directions of propagation. These configurations were covered in detail in Section 4.7.

8.7 Reflection Analysis

Multipath fading is caused predominantly by nonspecular ground reflections interfering with an attenuated main signal during ducting conditions. The reflection condition is not stable, hence the deep fading occurs for very short periods—typically milliseconds. Ground reflections become more serious if the whole wavefront is reflected in-phase, a so-called specular reflection. For this to happen the reflection plane needs to cover a sufficiently large area—typically the ground area covered by the first Fresnel zone. For the wavefront to be reflected in phase, the ground needs to be smooth compared with the wavelength of the signal. The divergence of the beam due to the Earth's curvature must also be taken into account. If the terrain variations are less than one-fourth of the wavelength the ground is considered to be smooth. A rough surface would cause the signal to scatter and the reflection could cause multipath fading but not a stable mean depression of the signal. At the higher microwave radio frequencies only a surface such as a body of water or flat land is smooth enough to cause a specular reflection.

The geometry of the path reflection is critical. When analyzing paths for reflections one needs to determine where the reflection point is. The fact that the path goes over water is not a guarantee that a reflection will occur. It can be seen from Figure 8.8 that the reflection point in case B is not on the water and therefore may not be a problem.

Physics dictates that at the point of reflection the angle of incidence and reflection must be equal. One needs to define the reflective plane and then identify the reflection points. This is done by drawing the tangent planes off the profile and considering which rays have the angle of incidence equal to the angle of reflection. The reflection angle is known as the grazing angle. The smaller the grazing angle, the smoother the surface appears—just as sunlight on a tar road can look like glass if viewed from a low enough angle.

The effect on the radio system also depends on the mean phase delay of the signal. A long delay can cause ISI. If the delay is short enough the reflected

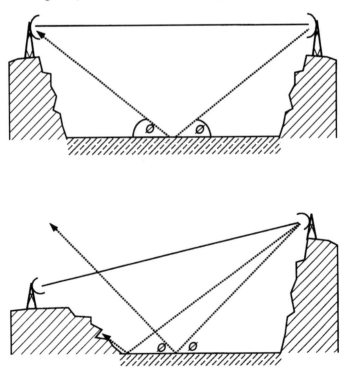

Figure 8.8 Reflection geometry.

signal may actually arrive within the symbol period of the signal and have a constructive interference effect. In narrowband systems the reflection causes a reduction in thermal fade margin. In wideband systems a reflection can have the same effect as dispersive fading. In other words, inband distortion can lead to errors when the delays are long unless the amplitude of the reflected signal is heavily attenuated by an obstruction, or divergence and roughness effects.

When doing a reflection analysis one must remember that the reflection point will change as the k factor changes. It is possible by careful positioning the antenna heights to minimize the effect of a reflection. The antenna position should be chosen to minimize the attenuation at the median k condition while ensuring that an acceptably small attenuation is present for low and high values of k. Where this is not achievable, space diversity may be required. The position of the diversity dish should be such that it experiences a good signal under the conditions where the main dish is experiencing a signal null. This should be checked for all expected values of k.

References

[1]   CCIR Report 718-3, Geneva, 1990.

[2]   Rummler, W. D., *A New Selective Fading Model: Application to Propagation Data*, Bell Laboratories, 1978.

[3]   CCIR Report 338-6, Geneva, 1990.

[4]   ITU-R P 530-7, Geneva, 1997.

[5]   ITU-R F.1093, Geneva, 1994.

[6]   ITU-R P.838, Geneva, 1992.

[7]   TR-TSY-000752/499, Bellcore Standard.

[8]   Fabbri, F., S. Giaconia, and L. F. Mojoli, *The Red Sea Microwave Project*, Telettra, June 1980.

[9]   Hartman, W. J., and D. Smith, Tilting Antennas to Reduce Line-of-Sight Microwave Fading, July 1976.

[10] ITU-R F.752-1, Geneva, 1994.

[11] Vigants, A., "Space Diversity Engineering," *Bell System Technical J.*, 1975.

9

Synchronous Digital Hierarchy

9.1 What Is Plesiochronous Digital Hierarchy?

Analog radio networks were based on frequency division multiplexing and hence called FDM systems. The digital systems that replaced them were based on time division multiplexing and use pulse code modulation (PCM) to form the primary digital line rate (E1 or T1). To create higher bit rates secondary multiplexers are used. This is not done synchronously but is made to look synchronous by a technique called stuffing. "Plesio" means nearly, hence the term plesiochronous (nearly synchronous) digital hierarchy (PDH).

Higher order multiplexers are used to bit interleave the incoming bit streams into a higher order stream. The multiplexers have to synchronize the incoming primary order streams so that they can be multiplexed into a single higher order bit stream. Each E1 or T1 stream is essentially free running because it is not locked to a central clock signal. The nominal E1 bit rate is 2048 kbit/s ±50 ppm. In a 2/8 secondary multiplexer, four 2-Mbit/s streams are multiplexed into an 8-Mbit/s stream as follows. The incoming bit streams are read into elastic store buffers using a clock that is extracted from the bit stream. The bits are written out each buffer one bit at a time and bit interleaved into the aggregate stream using the main multiplexer clock. To ensure that the fastest incoming stream does not cause the buffer to overflow, the multiplexer clock is run at a rate higher than the fastest incoming stream—in other words, 2048 kbit/s + 50 ppm (2048 102 b/s). There are also extra bits added to the secondary stream, so the clock rate needs to be even faster to allow the line clocks to be stopped while the extra overhead bits are added. Running the clocks faster means there is a natural tendency for the buffers to run empty. To avoid this, when a certain threshold is reached the buffer write clock is stopped and during this period "stuffing" bits are inserted into the aggregate signal. Special control bits are used to tell the demultiplexer at the other end which bits are real and which are stuffing so that the stuffing bits can be discarded. In addition to stuffing bits a frame alignment word (FAW) is added to create an overall frame of 8448 kbit/s. The higher rates of 34 Mbit/s and 140 Mbit/s are created in a similar way by multiplexing four of the lower rate signals. The so-called "mux-mountain" is shown in Figure 9.1.

In practice, double-step and triple-step multiplexers are used to skip the intermediate levels. A comparison on PDH rates defined in North America and Europe is presented in Table 9.1.

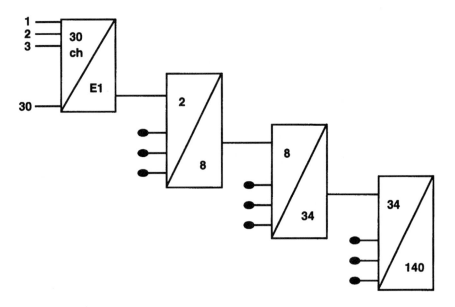

Figure 9.1 MUX mountain.

Table 9.1
Various PDH Standard Bit Rates

PDH (North America)	Bit Rate	PDH (ITU)	Bit Rate
T1 (DS 1)	1.544 Mbit/s	E1	2.048 Mbit/s
T2 (DS 2)	6.312 Mbit/s	E2	8.448 Mbit/s
T3 (DS 3)	44.736 Mbit/s	E3	34.368 Mbit/s
T4 (DS 4)	139.264 Mbit/s	E4	139.264 Mbit/s

9.2 Synchronous Networks (SDH/SONET)

With the demand for more bandwidth and the need for standardization, manageability, and flexibility in networks, a new standard was developed within the ITU. Work began around 1986 and in 1988 the first SDH standards were approved. The objective was to have one worldwide set of standards that would allow interoperability of different vendors' equipment within the same network. The standards were based on the North American SONET optical standard and were designed to ensure that the North American 1544-kbit/s and European 2048-kbit/s bit rates were both accommodated. The SDH standard uses a common bit rate of 155 Mbit/s. A comparison between the SONET and SDH rates is shown in Table 9.2.

9.2.1 What Is Synchronous Digital Hierarchy?

In PDH the lower rate signals are bit interleaved into the hierarchy, thus losing their original interface characteristics. Stuffing techniques are used to ensure that the overall signal can be demultiplexed at the distant end without requiring a common clock. In SDH the principle is to synchronously map the lower rate signal into a container that is thus embedded in the overall frame without losing its original interface characteristics. The container is kept synchronized to the frame using pointer techniques, which will be discussed later. An overhead is then added to the container and to the frame that allows manageability of the original signal right through the network. It is this aspect that allows SDH to have such powerful network management capability.

9.2.2 SDH Structures (Multiplexing)

In order to have a truly international standard the various existing PDH bit-rate interfaces must be accommodated in the SDH structure. This is done by allowing various interfaces to be mapped into the SDH frame, as shown by Figure 9.2 [1].

Table 9.2
Comparison of SDH and SONET Rates

SONET	Transport Level	Bit Rate	SDH
OC-1	STS-1	51.84 Mbit/s	STM-0
OC-2	STS-3	155.52 Mbit/s	STM-1
OC-12	STS-12	622.08 Mbit/s	STM-4
OC-48	STS-48	2488.32 Mbit/s	STM-16

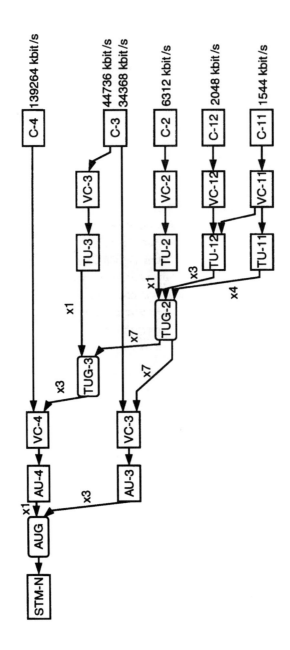

Figure 9.2 SDH mapping structure.

Consider the mapping of a 2-Mbit/s signal into the SDH frame. The original PDH signal will be at 2048 kbit/s with a 50 ppm variation. This is inserted into a container (C-12) where justification takes place using traditional stuffing techniques. This is done to compensate for the allowed frequency variations in the PDH and SDH bit rates. The bit stuffing ensures that the 50 ppm variation in the PDH signal does not result in errors when demapped at the final destination. The container is then placed into a virtual container (VC-12) where a path overhead is added. This overhead is carried with the signal throughout the network, even when cross-connected into a different SDH frame. It allows for maintenance and supervision of the signal throughout the network. It includes error detection, alarm indications, and a signal label. A pointer is then added to the virtual container to form a tributary unit (TU-12). This allows the SDH system to compensate for phase differences across the network or between networks. Three TUs are then multiplexed into a tributary unit group (TUG-2). Seven TUG-2s are multiplexed into one TUG-3. This is the same-sized unit that would be used to map, for example, an E3 signal into the SDH frame. Three TUG-3s are then multiplexed via the AU-4 and AUG into the STM-1 frame. The tributary bit rate contained within each virtual container is shown in Table 9.3.

9.2.2.1 SDH Frame Structure

The basic SDH frame consists of a matrix of 8-bit bytes organized into 270 columns and 9 rows. The frame duration is 125 μs. There are three main areas of interest:

1. Section overhead (SOH);
2. AU pointer;
3. Payload.

This is illustrated in Figure 9.3.

Table 9.3
Bit Rates Supported by Each Virtual Container

Virtual Container	Tributary Bit Rate
VC-11	T1 (1.544 Mbit/s)
VC-12	E1 (2.048 Mbit/s)
VC-3	E3 (34.368 Mbit/s) or T3 (44.736 Mbit/s)
VC-4	E4 (139.264 Mbit/s) or ATM (149.76 Mbit/s)

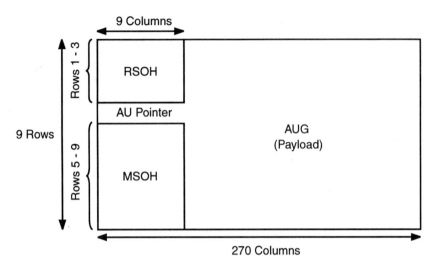

Figure 9.3 SDH Frame structure.

9.2.2.2. Section Overheads

The SOH is used for the individual transport system to allow error monitoring, alarm monitoring, and services and network management. It contains two parts: a regenerator section overhead (RSOH) and a multiplexer section overhead (MSOH). The RSOH is terminated at each regenerator, whereas the MSOH is only terminated at the multiplexer and is unaffected by the regenerator. This facilitates monitoring of the path between multiplexers separately to the individual regenerator sections. In the RSOH there are framing bytes (A1, A2), a regenerator data channel for management (D1 to D3), a regenerator order wire channel (E1), and a spare user channel (F1). In the MSOH there is a multiplexer data channel for management (D4 to D12), a multiplexer order wire channel (E2), multiplexer section protection switching (K1, K2), and error-monitoring bytes using bit interleaved parity (B1, B2). Recently a S_1 byte was defined for synchronization status.

9.2.2.3 Path Overheads

Each virtual container is only assembled and disassembled once. A path overhead is carried with the virtual container between different transport systems allowing end-to-end circuit monitoring. Two types of path overhead are defined: a high-order path overhead associated with VC-3 and VC-4 levels and a low-order path overhead associated with VC-2 and VC-12 levels. The high-order path overhead includes a path trace byte (J1), a signal label byte (C2), a path status byte (G1), and an error-monitoring byte (B3). The status of the high-order virtual container can thus be monitored across the network. The low-order path overhead is called the V5 byte and includes a bit for error monitoring (BIP-2), alarm bits (FEBE and FERF), and a three-bit signal label.

9.2.2.4 Pointers

A synchronous system relies on the fact that each clock is in phase and frequency synchronism with the next. In practice this is impossible to achieve; therefore, phase and frequency deviations will occur. Within a network the clock frequency is extracted from the line signal, however, phase variations can still occur from accumulated jitter over the network. At the network interface frequency variations can also occur. The way that SDH overcomes this problem is to use pointers to "point" to the address of the beginning of the virtual container within the frame. The AU-4 pointer shows where the VC-4 starts within the frame. Within the VC-4 are TU pointers that show where the lower order VCs such as the VC-12 start relative to the position of the VC-4. The AU-4 pointer is made up of three bytes H1 to H3. The actual pointer value is contained within H1 and H2 and H3 is reserved for negative justification. The actual pointer value is contained within 10 bits that have a maximum value of 782. Each increment in pointer value adjusts the addressing by three bytes in the frame. The initial pointer value corresponds to the phase difference between the arrival of the tributary and the empty tributary unit within the frame at the time the tributary is mapped into the virtual container. If the phase varies between the read and write clocks such that the terminating digital stream input buffers show a tendency to either overflow or run empty, a pointer adjustment will occur.

9.2.3 SDH Equipment

SDH equipment consists of four basic building blocks: a terminal multiplexer, an add-drop multiplexer (ADM), a cross-connect switch, and a regenerator. Since a regenerator is a fiber device, it will not be covered here.

9.2.3.1 Terminal Multiplexers

A terminal multiplexer is used to terminate a point-to-point SDH link. They typically operate at STM-4 and above and can terminate both SDH and PDH traffic. The aggregate signals are usually protected in a 1:1 or 1:n hardware arrangement.

9.2.3.2 Add-Drop Multiplexers

The ADM is a fundamental building block of an SDH network. It allows one to add and drop tributaries without demultiplexing the whole SDH signal. It is not quite as simple as "plucking out" a single 2-Mbit/s stream from the aggregate stream because a VC-4 needs to be broken down into its VC-12s before access to a 2-Mbit/s signal can be obtained; however, the principle is that tributaries can be extracted and added using software control. The ADMs are typically used for STM-1 and STM-4 capacities because at the STM-16 level and above, cross-connects are usually required. The aggregates are usually termed East and West. The tributaries are transmitted to both East and West directions, and on the receive side they can be selected from either the East or West direction via a software switch. This is shown in Figure 9.4.

When ADMs are configured in a ring this ability to automatically switch between East or West directions provides the self-healing resilience to failures.

9.2.3.3 Cross-Connect Switches

Digital cross-connect switches (DXC) are used to cross-connect traffic between aggregate streams. This allows grooming of the SDH traffic and powerful rerouting ability. Higher order cross-connects allow protection of failed circuits using an Automatic Network Protection System (ANPS). Cross-connects are classified according to the hierarchical trunk termination and tributary cross-connect level. For example, a DXC 4 /1 cross-connect will be capable of terminating a level 4 signal (STM-1 or E4) and cross-connecting at level 1 (low-order path such as VC-12). A DXC 4 /4 can terminate level 4 trunks (STM-1) but will be limited to cross-connect at level 4 as well (VC-4). In smaller networks two ADMs can be configured as a DXC.

9.2.4 SDH Networks

The core SDH networks are almost exclusively built with fiber optics due to the high bandwidth requirements. Radio is used in the access network and on

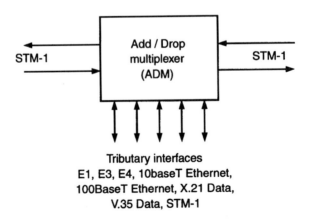

Figure 9.4 Add-drop multiplexer (ADM) with East/West aggregates.

some of the long-haul routes using $n \times$ STM-1 systems. These networks are built in ring and linear configurations.

9.2.4.1 Self-Healing Rings

It is wise when designing a transmission network to build a ring topology to allow alternate routing of circuits. A very resilient ring topology is when ADMs are configured in self-healing rings. The concept is to provide a pair of fiber optic cables between each ADM. The traffic on each fiber is configured to travel in opposite directions around the ring. The main traffic is fed, for example, to the clockwise fiber (the service (S) fiber); and the same traffic is also fed to the counterclockwise fiber, or the protection (P) fiber. The ADM normally switches the VC from the P fiber to the tributary output; however, if failure of that path occurs, it will automatically switch to the P fiber path within 50 ms, thus restoring the traffic. An SDH self-healing ring is shown in Figure 9.5.

Radio links can easily be used in these SDH networks by replacing the fiber connections with a point-to-point radio. Since the ring topology provides the equipment and path protection, the radios can be configured as nonprotected (1 + 0). If the radio path requires space diversity due to multipath fading problems, this will still be required in the self-healing ring because the ADM switching is not hitless.

9.2.4.2 Linear Routes

In many networks it is not possible to create self-healing rings and, therefore, SDH topologies must also support trunk routes and star networks. In this

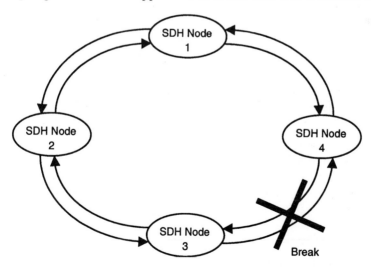

Figure 9.5 SDH ring with fiber cut.

case, since there is no ring protection, protected radios (Hotstandby) and duplicated ADM aggregate cards should be used. SDH trunk routes often require capacities in excess of STM-1, which present no problem to the virtually infinite bandwidth capability of fiber (especially with wave division multiplexing) but present difficulties on radio systems especially within the limited channel spacing available on existing frequency plans. Radio systems use $n + 1$ switching systems to provide $n \times$ STM-1 channels with one standby channel that is shared for protection. Often increased capacity is catered for using the same frequency on the alternate polarization using a dual-polar antenna. Fading effects such as rain can rotate the signal, thus reducing the cross-polar discrimination. Devices called cross-polar interference cancelers (XPICs) are used to dynamically keep the two polarized signals orthogonal to one another. The use of an XPIC can be avoided in some cases by running the cross-polar signal on an adjacent channel.

9.2.5 Synchronization

SDH systems may require eternal synchronization equipment to keep the accumulated jitter over a network within the limits set by the ITU. The ITU sets limits on jitter and wander in a network [2].

9.2.5.1 Why is Synchronizing Required?

It often seems to be a contradiction in terms that external synchronization is required in a SDH network. It is important to realize that synchronous in this context does not mean that the incoming tributaries are synchronized; it refers to the position of the virtual containers within the frame being located in known positions for multiplexing purposes. The pointer adjustments that solve the internal synchronization problem of the multiplexer causes a large amount of jitter on the tributaries. Each pointer adjustment shifts the tributary position by three bytes, causing a stepped phase jump of 24 unit intervals (UI) at a time. If the difference between the clocks at the perimeter of the network is significant, lots of pointer adjustments would occur and the jitter objectives would be exceeded. Some equipment may also not tolerate the stepped nature of the phase jitter as the pointer adjustment occurs. To avoid pointer movements one needs to ensure that the network element clocks are not allowed to drift too far from each other. To do this a hierarchy of clock sources has been defined. The hierarchy is defined by the ITU [3].

9.2.5.2 Timing Sources

A clock can be characterized by its inherent stability and its ability to remember its previous accuracy if it was previously slaved to a higher order clock. This latter mode is termed holdover. A synchronized clock is one that derives its accuracy from a higher level source. A free running clock is one that runs on its own internal oscillator. Three levels of clocks have been defined for transmission networks: A primary reference clock (PRF) [4]; a synchronization supply unit (SSU), secondary reference clock [5]; and a synchronous equipment clock (SEC) [6].

Primary Reference Clock

The ITU [4] specifies that a PRC should have an accuracy exceeding 1 part in 10^{11} compared with universal time coordinated (UTC), which is the international absolute time reference. This ensures that a network will only experience a controlled slip once every 70 days. To achieve this accuracy a Cesium clock is required. Alternatively, a GPS source can be used to synchronize a cheaper clock such as a Rubidium-based one, which can achieve an even higher overall accuracy than a Cesium-based one on its own. PRCs are required at main switching centers and should be used where one network connects with another network. In a GSM network a G.811 clock should be used at the MSC.

Secondary Reference Clock

The timing source derived from G.811 is transferred across the network, extracted off the aggregate SDH stream. After a number of network elements this signal will become impaired and the accumulated jitter should be filtered using a narrowband SSU using Rubidium to achieve the high-quality holdover. A transit and local clock reference is defined and specified by ITU G.812.

Synchronous Equipment Clock

Each network element has a built-in SDH equipment clock in terms of the internal crystal oscillators. This clock is usually locked to the incoming aggregate signal but should have reasonable holdover ability should the incoming stream be lost. The requirements of this clock are specified by the ITU [5].

9.2.5.3 Practical Advice

Small SDH networks will usually not require any external synchronization equipment, although a G.811 reference clock will still be required at the main switching center where connection to the PSTN switch is made—this is equally true of PDH networks. A new G.811 standard reference clock is required after 60 network elements, and after every 20 network elements an SSU should be used to filter the jitter to tolerable limits. One should be careful not to create timing loops where a network element tries to time itself off an element that is itself timed off that network element. The synchronization status message within the S1 byte is used to ensure that this does not happen by defining which inputs to use for synchronization and which ones not to use.

9.2.6 Benefits of SDH

Usually the benefits of SDH are listed at the beginning of the chapter. The typical benefits listed are common international standards, simpler multiplexing (no MUX mountain), and embedded network management. While these benefits are real, this discussion has purposely been left to the end of the chapter in order to be able to meaningfully discuss the real benefits and trade-offs.

First, although it is true that the SDH standard does cater to international interfaces, the SONET standard is coexisting rather than being replaced by SDH. It is unlikely that this will change in the near future due to the explosive growth of SONET networks in Northern America. Standards do exist for the basic transport rates especially at STM-1 and above, but standardization is slow for sub-STM-1 and sub-sub-STM-1, which are of interest to radio networks operating in the access portion of the telecommunications network. Interoperability of synchronous network elements has been proved in America in various SONET interoperability forums, but true intervendor operation is still a long way off in SDH. The network management Q3 interface that should allow different vendors' equipment to be fully integrated and managed by a common management platform is still not fully defined.

Another advantage that SDH is reported to have is that it removes the MUX mountain problem of PDH. With modern PDH equipment, this can no longer be considered a major factor. PDH multiplexers have double- and triple-step multiplexing integrated into a single shelf with less complexity than an ADM.

The embedded network management capability is not without added complexity and cost. In fact, at the primary level, the basic traffic-carrying ability of SDH is reduced compared to PDH due to huge bandwidth overhead required for management. More than 5 Mbit/s is allocated for overheads; so in a 155-Mbit/s (STM-1) signal 63 E1 streams are accommodated, yet a comparable PDH 140-Mbit/s system can carry 64 E1 streams. The cost and complexity of ADM equipment compared to a comparable PDH system is significant, especially when one takes into account the extra synchronization equipment required. Is PDH therefore a better alternative?

The answer is a resounding NO! The reason has to do with the way that telecommunication services have changed. With deregulation of the telecommunications market, service operators have a choice of transmission networks. It is essential for the transmission operator to be able to monitor and control the end-to-end circuits in the network and measure their quality. Flexibility of the circuits provided is absolutely key to providing a modern day service. The days of waiting for technicians to drive out to a site to make network changes are over. It is essential to be able to make these changes quickly and efficiently without a large pool of highly trained technicians. SDH allows network changes to be done from the network management center. With the high cost of bandwidth in a cost-competitive market, bandwidth on demand is becoming the trend. SDH is ideally suited to meeting these network needs. It is also future proof against increased capacity requirements and supports standards such as ATM. SDH is no longer a new and untried technology. It is well established and the obvious choice for any network exceeding a capacity of around 34 Mbit/s. By extending the concept of using virtual containers to provide manageability of circuits right to the end user, it is likely that SDH systems on radio will migrate below the STM-1 level, thus providing SDH networks within the access network right to the customer premises.

References

[1]   ITU-T G.707, Geneva, 1996.

[2]   ITU-T G.822, Geneva, 1988.

[3]   ITU-T G.803, Geneva, 1997.

[4]   ITU-T G.811, Geneva, 1997.

[5]   ITU-T G.812, Geneva, 1998.

[6]   ITU-T G.813, Geneva, 1996.

List of Acronyms and Abbreviations

AC	Alternating current
ADM	Add drop multiplexer
ADPCM	Adaptive differential pulse code modulation
AGC	Automatical gain control
ALC	Automatic level control
AM	Amplitude modulated
ATM	Asynchronous transfer mode
ATPC	Automatic transmit power control
BPSK	Bipolar phase shift keying
BER	Bit error ratio
BSC	Base station controller
BTS	Base transceiver station
C/I	Carrier to interference ratio
CCIR	Comite Consultatif International de Radio
CCITT	Comite Consultatif International Telephonique et Telegraphique
CODEC	Coder decoder
DPSK	Differential phase shift keying
dB	decibels
DC	Direct current
DECT	Digital enhanced cordless telecommunications
DFM	Dispersive fade margin
DM	Degraded minutes

DTM / DEM	Digital terrain (Elevation) model
DTMF	Dual tone multi-frequency
DXC	Digital cross connect
EBR	Errored block ratio
EIRP	Effective isotropic radiated power
EM	Electromagnetic
EMC	Electromagnetic compatibility
EOW	Engineering order wire
ES	Errored seconds
ESR	Errored second ratio
ETSI	European Telecommunications Standards Institute
FAW	Frame alignment word
F/B	Front to back ratio
FD	Frequency diversity
FEC	Forward error correction
FFM	Flat fade margin
FM	Frequency modulated
FSK	Frequency shift keying
FSL	Free space loss
GHz	Gigahertz (10^9)
GPS	Global positioning system
GSM	Groupe Speciale Mobile - Global System Mobile
HF	High frequency
HSB	Hot standby
HTML	Hypertext markup language
IDU	Indoor unit
IF	Interference frequency
IFRB	International Frequency Regulations Board
IMP	Intermodulation product
IRF	Interference reduction factor
IP	Internet protocol
ISDN	Integrated services digital network
ITU	International Telecommunications Union
ITU-R	ITU – Radiocommunications Agency
ITU-T	ITU – Telecommunications Agency
KHz	Kilohertz (10^3)
LAD	Linear amplitude distortion

LO	Local oscillator
MHz	Megahertz (10^6)
MODEM	Modulator - Demodulator
MSC	Mobile switching center
MSOH	Multiplexer section overhead
MTTR	Mean time to repair
MULDEM	Multiplexer - Demultiplexer
MUX	Multiplexer
NFD	Net filter discrimination
nm	Nanometer (10^{-9})
ODU	Outdoor unit
OOK	On-off keying
PCM	Pulse code modulation
PCN /PCS	Personal communications networks/systems
PDH	Plesiochronous digital hierarchy
PLL	Phase locked loop
pm	Picometer (10^{-12})
PRC	Primary reference clock
PSK	Phase shift keying
PSTN	Public switched telephony network
PTO / PTT	Public telecommunications operator
QAM	Quadrature amplitude modulation
QPSK	Quadrature phase shift keying
RBER	Residual bit error ratio
RF	Radio frequency
RL	Return loss
ROI	Return on investment
RPE	Radiation pattern envelope
RRB	Radio Regulations Board
RSOH	Regenerator section overhead
RX	Receiver
SD	Space diversity
SDH	Synchronous digital hierarchy
SDP	Severely disturbed period
SEC	Synchronous equipment clock
SES	Severely errored seconds
SESR	Severely errored second ratio

SONET	Synchronous optical networks
SSU	Synchronization supply unit
STM-	Synchronous transport mode
SWR	Standing wave ratio
TAE	Transversal adaptive equalizer
T/I	Threshold to interference ratio
TE	Transverse electrical
TM	Transverse magnetic
TEM	Transverse electromagnetic
TETRA	Terrestrial trunked radio standard
THz	Terahertz (10^{12})
TU	Tributary unit
TUG	Tributary unit group
TX	Transmitter
UHF	Ultra high frequency
UI	Unit intervals
UPS	Uninterruptable power supply
UV	Ultraviolet
VC	Virtual container
VCO	Voltage controlled oscillator
VHF	Very high frequency
VSWR	Voltage standing wave ratio
WARC	World Administrative Radio Conference
WDM	Wave division multiplexing
WLL	Wireless local loop
XPD	Cross-polar discrimination
XPIC	Cross polar interference canceller
XTE	Cross track error

About the Author

Trevor Manning graduated in 1984 from the University of Natal (Durban, South Africa) with a B.Sc. degree in electronic engineering. After graduation he worked with the South African national electricity utility (ESKOM). This company owns and operates a countrywide telecommunications network comprised of in excess of 800 sites. Mr. Manning was involved in radio system design and site selection for the SDH digital upgrade for ESKOM's $100 million radio network. In 1991 he was seconded to work at Telettra based in Vimercate (Italy) where he worked alongside world-renowned propagation expert Umberto Casiraghi. Before leaving ESKOM he occupied the post of Chief Engineer and was responsible for the national microwave radio system. He has traveled extensively around the world visiting various experts in the telecommunications field and has presented a number of technical papers at various conferences. Mr. Manning is currently employed by Digital Microwave Corporation (DMC) as technical director within the marketing department of the Europe, Middle East, and Africa sector based in Coventry (United Kingdom). He is married with three children, and his interests include music, playing tennis, and squash.

Index

223

Recent Titles in the Artech House Microwave Library

Advanced Automated Smith Chart Software and User's Manual,
Version 3.0, Leonard M. Schwab

*C/NL2 for Windows: Linear and Nonlinear Microwave Circuit Analysis
and Optimization, Software and User's Manual*, Stephen A. Maas
and Arthur Nichols

Computer-Aided Analysis of Nonlinear Microwave Circuits,
Paulo J. C. Rodrigues

Design of FET Frequency Multipliers and Harmonic Oscillators,
Edmar Camargo

Design of RF and Microwave Amplifiers and Oscillators,
Pieter L. D. Abrie

Feedforward Linear Power Amplifiers, Nick Pothecary

*FINCAD: Fin-Line Analysis and Synthesis for Millimeter-Wave
Application Software and User's Manual,* S. Raghu Kumar,
Anita Sarraf, and R. Sathyavageeswaran

Generalized Filter Design by Computer Optimization,
Djuradj Budimir

GSPICE for Windows, Sigcad Ltd.

Introduction to Microelectromechanical (MEM) Microwave Systems,
Hector J. De Los Santos

*LINPAR for Windows: Matrix Parameters for Multiconductor
Transmission Lines, Software and User's Manual, Version 2.0,*
Antonije Djordjevic *et al.*

Microwave Engineers' Handbook, Two Volumes, Theodore Saad,
editor

*Microwave Filters, Impedance-Matching Networks, and Coupling
Structures,* George L. Matthaei, Leo Young, E.M.T. Jones

Microwave Mixers, Second Edition, Stephen Maas

Microwaves and Wireless Simplified, Thomas S. Laverghetta

Microwave Radio Transmission Design Guide, Trevor Manning

MULTLIN for Windows: Circuit-Analysis Models for Multiconductor Transmssion Lines, Software and User's Manual, Antonije R. Djordjevic *et al.*

The RF and Microwave Circuit Design Handbook, Stephen A. Maas

RF Design Guide: Systems, Circuits, and Equations, Peter Vizmuller

RF and Microwave Coupled-Line Circuits, Rajesh Mongia, Inder Bahl, and Prakash Bhartia

RF Power Amplifiers for Wireless Communications, Steve C. Cripps

RF Systems, Components, and Circuits Handbook, Ferril Losee

SPURPLOT: Mixer Spurious-Response Analysis with Tunable Filtering, Software and User's Manual, Version 2.0, Robert Kyle

TRANSLIN: Transmission Line Analysis and Design, Software and User's Manual, Paolo Delmastro

TRAVIS Pro: Transmission Line Visualization Software and User's Manual, Professional Version, Robert G. Kaires and Barton T. Hickman

For further information on these and other Artech House titles, including previously considered out-of-print books now available through our In-Print-Forever® (IPF®) program, contact:

Artech House
685 Canton Street
Norwood, MA 02062
Phone: 781-769-9750
Fax: 781-769-6334
e-mail: artech@artechhouse.com

Artech House
46 Gillingham Street
London SW1V 1AH UK
Phone: +44 (0)20-7596-8750
Fax: +44 (0)20-7630-0166
e-mail: artech-uk@artechhouse.com

Find us on the World Wide Web at: www.artechhouse.com